Palgrave Series in Asia and Pacific Studies

Series Editors
May Tan-Mullins
University of Nottingham Ningbo China
Ningbo, Zhejiang, China

Adam Knee
Lasalle College of the Arts
Singapore

The Asia and Pacific regions, with a population of nearly three billion people, are of critical importance to global observers, academics, and citizenry due to their rising influence in the global political economy as well as traditional and nontraditional security issues. Any changes to the domestic and regional political, social, economic, and environmental systems will inevitably have great impacts on global security and governance structures. At the same time, Asia and the Pacific have also emerged as a globally influential, trend-setting force in a range of cultural arenas. The remit of this book series is broadly defined, in terms of topics and academic disciplines. We invite research monographs on a wide range of topics focused on Asia and the Pacific. In addition, the series is also interested in manuscripts pertaining to pedagogies and research methods, for both undergraduate and postgraduate levels. Published by Palgrave Macmillan, in collaboration with the Institute of Asia and Pacific Studies, UNNC.

More information about this series at
http://www.palgrave.com/gp/series/14665

Wu Deng • Ali Cheshmehzangi

Eco-development in China

Cities, Communities and Buildings

Wu Deng
University of Nottingham
Ningbo China
Ningbo, Zhejiang, China

Ali Cheshmehzangi
University of Nottingham
Ningbo China
Ningbo, Zhejiang, China

Palgrave Series in Asia and Pacific Studies
ISBN 978-981-10-8344-0   ISBN 978-981-10-8345-7 (eBook)
https://doi.org/10.1007/978-981-10-8345-7

Library of Congress Control Number: 2018940466

© The Editor(s) (if applicable) and The Author(s) 2018
This work is subject to copyright. All rights are solely and exclusively licensed by the Publisher, whether the whole or part of the material is concerned, specifically the rights of translation, reprinting, reuse of illustrations, recitation, broadcasting, reproduction on microfilms or in any other physical way, and transmission or information storage and retrieval, electronic adaptation, computer software, or by similar or dissimilar methodology now known or hereafter developed.
The use of general descriptive names, registered names, trademarks, service marks, etc. in this publication does not imply, even in the absence of a specific statement, that such names are exempt from the relevant protective laws and regulations and therefore free for general use.
The publisher, the authors, and the editors are safe to assume that the advice and information in this book are believed to be true and accurate at the date of publication. Neither the publisher nor the authors or the editors give a warranty, express or implied, with respect to the material contained herein or for any errors or omissions that may have been made. The publisher remains neutral with regard to jurisdictional claims in published maps and institutional affiliations.

Cover credit: Ali Cheshmehzangi

Printed on acid-free paper

This Palgrave Macmillan imprint is published by the registered company Springer Nature Singapore Pte Ltd. part of Springer Nature.
The registered company address is: 152 Beach Road, #21-01/04 Gateway East, Singapore 189721, Singapore

We would like to collectively dedicate this work to the future eco-development of China. What is being done in China will profoundly impact the global endeavour moving towards a sustainable built environment.
Wu and Ali

Wu Deng would like to dedicate this work to a poem that has inspired him in these years:
There is a pleasure in the pathless woods (无径之林,常有情趣)
There is a rapture on the lonely shore (孤独之岸,亦有欢欣)
There is society where none intrudes (心中桃源,何人能识)
George Gordon Byron (1788–1824)

Ali Cheshmehzangi dedicates this book to two great ladies, Afsaneh and Juan.

Preface

The rapid expansion of urbanisation has inevitably brought severe pressures on resource conservation and environmental protection in Chinese cities. Some major matters, such as increasing pollution, traffic and energy consumption in the urban areas, are already causing concern to city managers in China. There is an urgent need to respond effectively to such increasing problems. In recent years, the country has initiated policies, strategies and pilot projects at both national and local levels to address these issues. These are often initiated with the purpose of identifying an urban development model that can be replicable in a wide range of contexts. After several years of progress, China now appears to be at the forefront of reshaping and redeveloping the urban environment. There is a growing interest in providing or constructing eco-development at various spatial levels of the built environment.

This work takes a look at the contemporary eco-development projects in China at three different, but overlapping spatial scales—Macro (City Level), Meso (Community/Neighbourhood Level) and Micro (Building Level). A selection of 21 case study projects at different spatial levels are reported on and evaluated. Many of these case studies will be unfamiliar to readers outside China. Therefore, this book is also aimed to provide international readers with a comprehensive and timely picture of current eco-development practice and its outlook in China. Furthermore, the book argues that potentially greater performance gains lie in synergies between these spatial scales. An integrated model that addresses the three interwoven spatial levels will be beneficial to future eco-development in China and anywhere around the globe.

The work includes both educational and practical features that are crucial for the current phase of development in China. It aims to identify the gaps and provide strategies and solutions for future eco-development that is expected to take place in China in the coming decades, and supply useful references for eco-development in other countries. This will be a strong foundation for studies in the fields of urbanism, sustainable development, eco-design, urban geography and development studies.

This book will be of interest to a variety of readers, from practitioners and academics who are working on urban planning, architecture, urban geography, China studies, urban sustainability, environmental sciences and development studies, to government agencies who plan to promote eco-development in their cities. It should also serve the needs of planning and architecture students who wish to understand current eco-development practices and strategies at city, neighbourhood/community and building levels.

It is hoped that readers will find in this book an exciting new dimension to the appreciation of urban development in China. Understanding what is happening now in China can bring a new vision to future eco-development, and can also provide us with a visionary platform to future trends and possibilities in other contexts.

Ningbo, Zhejiang, China   Wu Deng
Ningbo, Zhejiang, China   Ali Cheshmehzangi

Acknowledgements

The authors have obtained a funding grant from the Ningbo New Eastern City Administrative Commission to conduct research on eco-city and green building development. This grant allows us to travel to the case study projects and hire student assistants for drawings and data collection. Prior to this, the authors have also benefitted from a variety of eco-city research projects funded by the Faculty of Science and Engineering at the University of Nottingham Ningbo China. The case study selection of eco-city projects also benefitted from the horizon scanning report, *Smart-Eco Cities in China: Trends and City Profiles 2016*, which was conducted under a separately funded project between 2015 and 2018. In this regard, Ali Cheshmehzangi would like to thank and acknowledge the National Natural Science Foundation of China (NSFC) for *project number 71461137005*. These funded projects collectively allowed the authors to work closely with various stakeholders of some of the case studies across China.

We would like to thank those PhD students—Xavier Francis Ochieng, Felix Ikworia Osebor, Linjun Xie, and Bamidele Akinwolemiwa—for providing supporting materials and drafting some of the case studies. Our thanks also go to our undergraduate students—Sicheng Chen, Yintao Lv, Hui Shi and Yiyi Chen—for helping with drawings and reference checking. The authors would like to thank Dr John Blair for his editorial help during the book proposal preparation. Finally, the authors would also to thank the series editors for their support throughout the process.

Contents

1 **Introduction**   1
    1.1   *A Historic Perspective*   1
    1.2   *A Glimpse of the Book*   6
    References   10

2 **Sustainability and Development: Challenges, Implications and Actor Constellations**   13
    2.1   *Introduction*   13
    2.2   *Urban Sustainability in Context*   17
       2.2.1   *Setting the Scene*   17
       2.2.2   *Challenges of the Sustainable City*   19
       2.2.3   *Eco-City: A Newly Emerged Global Phenomenon*   21
    2.3   *Dimensions and Spatial Levels of Urban Sustainability*   24
       2.3.1   *Understanding the Built Environment*   24
       2.3.2   *Dimensions of Urban Sustainability*   26
       2.3.3   *Spatial Levels of Urban Sustainability*   28
       2.3.4   *Integrated Thinking of Urban Sustainability*   36
    2.4   *Actor Constellations of Sustainability*   40
       2.4.1   *Urban Sustainability: Actor Constellations*   40
       2.4.2   *Eco-Development: Actors and Barriers*   43
    References   47

3 Eco-Development in the Global Context — 51
- 3.1 Common Threads from the Global Examples — 52
- 3.2 Global Examples at City Level — 54
  - 3.2.1 Freiburg, Germany — 54
  - 3.2.2 Curitiba, Brazil — 55
- 3.3 Global Examples at the Neighbourhood Level — 57
  - 3.3.1 Beddington Zero Energy Development, London, UK — 57
  - 3.3.2 Hammarby Sjöstad, Stockholm, Sweden — 58
- 3.4 Global Examples at the Building Level — 60
  - 3.4.1 The Crystal, London, the UK — 60
  - 3.4.2 Passive House in Germany — 62
- 3.5 Evaluating the Sustainability Performance of the Built Environment — 65
  - 3.5.1 Green Building Evaluation — 65
  - 3.5.2 Neighbourhood Sustainability Rating Systems — 68
  - 3.5.3 Sustainable City Rating Systems — 70
- 3.6 How Do Global Developments Inform China? — 74
- References — 77

4 Eco-Development in the Chinese Context — 81
- 4.1 China's Urbanisation Era — 81
- 4.2 Towards a Comprehensive Development Model — 84
  - 4.2.1 China's Challenges of Urban Growth and Urbanisation: Lessons for the Future — 84
  - 4.2.2 Urbanism and Urbanisation: Towards a Comprehensive Development? — 85
- 4.3 Evolution of Eco-Development in the Chinese Context — 87
  - 4.3.1 Policies on Eco-Development in the National Five-Year Plans — 88
  - 4.3.2 Current Incentives for Eco-Development in China — 94
- 4.4 Local Efforts for Eco-Development in China — 97
  - 4.4.1 Motivations from Local Governments — 97
  - 4.4.2 Local Initiatives: A Case Study of Ningbo — 99
- References — 102

5 Macro Level: Eco-City Cases in China — 105
- 5.1 Introduction — 105
- 5.2 Chongming Eco-Island, Shanghai — 106
- 5.3 Guiyang City, Guizhou — 115

5.4	Meixi Lake Eco-City, Changsha, Hunan	120
5.5	Guangming Eco-City, Shenzhen, Guangdong	125
5.6	Sino-Singaporean Tianjin Eco-City (SSTEC), Tianjin	131
5.7	Sino-Swedish Wuxi Eco-City, Wuxi, Jiangsu	143
5.8	Conclusions	149
References		152

6 Meso Level: Eco-Neighbourhood/Community Cases in China — 157

6.1	Introduction	157
6.2	The Sino-Singaporean Suzhou Industrial Park (SIP), Suzhou, Jiangsu	159
6.3	The Sino-German Qingdao Eco-Industrial Park, Qingdao, Shandong	162
6.4	Tengtou Village, Ningbo, Zhejiang	164
6.5	Gubeikou Township, Beijing	168
6.6	The Sino-Singaporean Guangzhou Knowledge City: Ascendas OneHub Business Park, Guangzhou, Guangdong	171
6.7	Hongqiao Business Park, Shanghai	173
6.8	The Beijing Olympic Village, Beijing	176
6.9	The Grand MOMA, Beijing	178
6.10	Conclusions	180
References		182

7 Micro Level: Green Building Cases in China — 187

7.1	Introduction	187
	7.1.1 Current Green Building Practice in China	187
	7.1.2 Passive Design	190
	7.1.3 Design for Green Building Certification	190
7.2	Center for Sustainable Energy Technologies (CSET), Ningbo, Zhejiang	191
7.3	Bruck Residence, Huzhou, Zhejiang	195
7.4	Sino-Italian Ecological and Energy-Efficient Building (SIEEB), Beijing	198
7.5	Shanghai Tower, Shanghai	200
7.6	Hangzhou Low-Carbon Science and Technology Museum, Hangzhou, Zhejiang	201
7.7	NanHai@COOL, Shenzhen, Guangdong	204

	7.8	Binghai Xiaowai Primary School, Tianjin	208
	7.9	Conclusions	209
	References	211

8 Future Eco-Development in China and Beyond 215
- 8.1 Introduction — 215
- 8.2 Past and Current Trends of Eco-Development at Three Spatial Levels — 217
  - 8.2.1 City Level/Macro — 217
  - 8.2.2 Neighbourhood/Community Level: Meso — 220
  - 8.2.3 Building Level/Micro — 223
- 8.3 Benefits of 'Eco' from a Multiscalar Perspective — 226
  - 8.3.1 Urban Resilience to Climate Change — 226
  - 8.3.2 The Regional Context — 227
  - 8.3.3 Stakeholder Constellations — 229
  - 8.3.4 Traditions — 232
  - 8.3.5 Breaking Misperceptions — 236
- 8.4 Prognoses for the Future — 237
  - 8.4.1 Emerging Trends — 237
  - 8.4.2 Burgeoning Building Technologies — 239
  - 8.4.3 Zero Energy Building and Life Cycle Consideration — 243
  - 8.4.4 Nature-Based Solutions — 245
  - 8.4.5 Integration: Concepts of Eco-Innovation and Eco-Fusion — 247
- References — 249

9 Concluding Remarks 255
- 9.1 Recapitulation — 255
- 9.2 SWOT analysis — 257
  - 9.2.1 Strengths — 257
  - 9.2.2 Weaknesses — 259
  - 9.2.3 Opportunities — 261
  - 9.2.4 Threats — 262
- 9.3 Final Remarks — 263
- References — 264

Index — 267

List of Figures

Fig. 1.1	Rocinha, the largest favela in Rio de Janeiro, Brazil. Source: UN-DESA (2014, p. 32) World Urbanisation Prospects	8
Fig. 2.1	Three sustainability pillars highlighted in the Brundtland Report (drawn by the authors)	15
Fig. 2.2	The interplay between the issues and spatial levels of the built environment, at three levels of 1. Building level (B), 2. Neighbourhood level (N), and 3. City Level (C). The issues are divided into four at the building level (namely: B1. Standards, B2. Technologies, B3. Installations, and B4. Users), three at the neighbourhood/community level (namely: N1. Patterns, N2. Public Open Space, and N3. Social Cohesion), and four dimensions at the city level (namely: C1. Policy, C2. Infrastructure, C3. Governance, and C4. Planning). Source: Drawn by the authors	38
Fig. 2.3	Green building development ecosystem—key barriers through the whole building life cycle. Source: Deng et al. (2016) (drawn by one of the authors)	44
Fig. 3.1	The working of a passive house (redrawn by the authors based on the original building plan)	64
Fig. 5.1	Overview of Chongming Eco-island plan (adapted and redrawn by the authors based on the most recent masterplan strategy development of Chongming Eco-island provided by the local government in the project exhibition hall)	109
Fig. 5.2	Roadmap and targets to be achieved by 2020 for constructing ecological civilisation in Guiyang. Source: Adapted and redrawn from Guiyang Evening News, http://wb.gywb.cn/epaper/gywb/html/2017-10/01/content_32894.htm	118

Fig. 5.3	The multiple layers of green strategy for green spaces, green corridor, and key spatial qualities of Meixi Lake Eco-City (adapted and redrawn by the Authors, from the KPF masterplan documents)	123
Fig. 5.4	Overview of Guangming Eco-City plan (adapted and redrawn by the authors based on the most recent masterplan strategy development of Guangming Eco-city)	128
Fig. 5.5	The neighbour layout in SSTEC indicating the concept of 'eco-cell' (adapted and redrawn by the authors, based on the SSTEC documents from the project website)	135
Fig. 5.6	The breakdown of KPIs in SSTEC (drawn by the authors)	137
Fig. 5.7	Conceptual model of the Sino-Swedish Wuxi Eco-City (adapted and redrawn by the authors based on the Swedish counterparts 2013)	144
Fig. 5.8	Layout of SSWEC/Wuxi Taihu New Town (adapted and redrawn by the authors based on the Swedish counterparts 2013)	145
Fig. 7.1	Centre for Sustainable Energy Technologies (CSET), UNNC (Redrawn by the authors from the original documents provided by the CSET administrations)	193
Fig. 7.2	Green systems employed in the Bruck Residence, Huzhou China (adapted and redrawn by the authors from the original system design)	196
Fig. 7.3	Green systems employed in the Hangzhou Low Carbon Science and Technology Museum Building (images taken and adapted by the authors)	199
Fig. 7.4	The project floor plan (redrawn by the authors based on the original plan provided by Professor Yu Liu from the Northwest University in China)	205
Fig. 9.1	The view of several shared bikes from various companies left outside the entrance area of the University of Nottingham Ningbo China. These bikes have now been banned from the university campus area (photo taken by the authors)	260
Fig. 9.2	Breakdown of urban key performance targets from urban level to community and building levels (image drawn by the authors)	263

LIST OF TABLES

Table 2.1	A matrix to highlight the positioning of sustainability dimensions across the three spatial levels	37
Table 2.2	Breakdown of the key issues at each spatial level of the built environment	37
Table 9.1	Summary of pathways to sustainable built environment from the consideration of all three spatial levels in the built environment	262

CHAPTER 1

Introduction

1.1 A Historic Perspective

For the first time in human history more than half of the world's population live in urban settlements. This currently accounts for about four billion people and is still increasing steadily. Some recent data do indicate slower population growth in comparison to the recent past, as a result of declining rates of fertility, mainly in Europe, North America and Asia (UN DESA 2015). An earlier projection indicates that most of the growing population is expected to be concentrated in urban areas (UN DESA 2014), particularly in the developing countries of Asia, Africa and South America.

From a historic perspective, the exponential growth in the urban population is a result of cities becoming the ultimate destination to fulfil the aspirations of material comfort, poverty reduction, prosperity and various opportunities. Yet, if such growth is uncontrolled and unplanned, the reverse effect is often the case, with cities being characterised by lower levels of material comfort, greater levels of poverty and increased slum development. Nevertheless, these characteristics have driven profound changes in the spatial distribution of population in most countries of the world, resulting in the growth of a very large number of urban agglomerations.

To successfully accommodate this level of population growth in cities, we require hindsight from a range of historical precedents. Can we gain

© The Author(s) 2018
W. Deng, A. Cheshmehzangi, *Eco-development in China*,
Palgrave Series in Asia and Pacific Studies,
https://doi.org/10.1007/978-981-10-8345-7_1

pointers from civilisations of the past, most of which have failed? To answer this question, we first take a brief look at the history of two renowned urban centres: one global and one in the context of China. The first example is the Sumerian city of Ur, a settlement from almost 4000 years ago, and which is now part of present-day Iraq. The city was the largest in the world from 2030 to 1980 BCE with a population of about 65,000 people (Rosenburg 2016). It sat proudly on the banks of the River Euphrates, at the hub of an extensive system of irrigation canals (Lilienfeld and Rathje 1998). The residents of the city had valuable environmental instincts, ardently recycling old utensils made from metal, weapons, anything made from wood given the aridity of that region, shards of broken pottery and 'even aging grand temples and palaces were turned into new variations of the buildings' (Lilienfeld and Rathje 1998, p. 11). Strangely, as the residents of Ur were compulsive recyclers, its bureaucrats instituted agricultural practices that led to declining fertility and lower crop yields. Traditional farming practices were to let fields lie fallow, which allowed salinity to diminish and nutrients, especially nitrogen, to regenerate in the soil. The bureaucrats' agricultural rules banned fallowing, since increases in food production were needed to keep pace with the growth in the numbers of merchants, artisans, soldiers and labourers. Many of the latter were brought in from rural areas to toil on new temples, palaces and other monuments of conspicuous consumption being built to 'memorialise the city's vast hubris' (ibid.).

The decrease in the time given to fallowing led to a rising water table, and lower levels of soil nutrients. At the same time, salt loads in the water brought from mountains in the historical region of Mesopotamia resulted in a sharp decline in crop yields (Lilienfeld and Rathje 1998). At this point the sensible reaction would have been to return for the city government to order a return to traditional farming practices. However, the bureaucrats simply demanded more from the farmers and diverted attention from the underlying problem by organising more labourers to build outrageous displays of the city's economic and political success. The farmlands were unable to support the demands of the city bureaucrats. Eventually, people starved and Ur became weaker. The city was challenged by long-term drought and warriors attacking from the north; and it eventually collapsed around 2000 BC.

The second historical example we consider is that of Anyang, the capital city of the Shang dynasty in China at around 1200 BC. Archaeological discoveries indicate that the city was large, with an area of around

15–24 km². Anyang was home to around 230,000 residents, most of whom were artisans and other workers in various workshops and industries (Marks 2012, pp. 43–47). Similar to the situation of Ur, more food was required to support an increasing population. Food production was available only from surrounding hinterlands as the land inside the city was occupied mainly by palaces, temples, altars, houses, factories and tombs. Accordingly, much of the surrounding forests were cleared for farmlands. The appetite for farming and farmland was so strong that the Shang people invaded the territory of neighbouring states to seize uncultivated land or turned hunting areas and grazing pastures into farmlands.

Unsurprisingly, much of the fate of the Shang state related to issues around agriculture and the annual harvest. With increased areas of cultivated farmland, agricultural yields were increased gradually and there was a significant increase in population. Shang people experienced a favourable climate and a fecund environment (Keightley 1999). Optimistic opinions were spread based on a belief that the natural world was in a time of climatic stability. They were, accordingly, unprepared when the climate change eventually arrived.

The North China plain was considerably warmer and wetter at this time, which helped to increase agricultural output. However, climatologists have thus far cannot identify the reasoning behind the climatic conditions of the time (around 1100 BC), which led to the region suddenly becoming colder and drier. Colder weather lowered harvest yields by shortening the growing season (Marks 2012, p. 51). A complete failure of the harvest would have caused food shortages not only for the villagers, but also for the royal elite, soldiers and foundry workers. Declining food supplies also led to fewer births, and hence to a falling population. As a result, farmers fled to the woods or other states, further decreasing the labour and military force available to the Shang royalty (ibid.). Around 1050 BC, the Shang dynasty fell, having been overrun by the Zhou state. The foolishness of the Shang's assumption of climate stability has many echoes of the modern world's attitude towards climate change.

Unfortunately, Ur and Anyang were not the only ancient settlements to act in such a foolhardy manner, with many polities suffering from the lust for 'quick profits rather than long term stability' (Lilienfeld and Rathje 1998, p. 13). It was not simply aggrandisement, but a combination of a fragile environment, unpredictable local climate conditions, and agricultural management techniques that led to a shortfall of essential foodstuffs in relation to a burgeoning population. Although advanced techniques for improving yield were adopted, a drive towards excessive demand without considering the natural capacity and a lack of preparation for climate change still led to the failure of Ur and Shang. With three millennia of hindsight, these actions appear to be imprudent in the extreme. However, they are quite representative of most of the civilisations of world history (Diamond 2011). We only picture two of the well-known examples to remind us of our recurring mistakes.

Similar to Ur and Anyang, today's cities offer greater geographic mobility, economic vitality and increased life expectancy. Present-day cities concentrate much of any country's economic activity, government services, commerce and transportation. Urban living is usually associated with higher levels of literacy and education, improved access to a more diversified labour market, and enhanced opportunities for cultural and political participation (UN DESA 2014). However, as the world continues to urbanise, environmental challenges have been increasingly concentrated in cities, particularly in countries where infrastructure is underdeveloped and policies are not implemented to ensure that the benefits of urban living are shared equitably. Inefficient building stock, high dependency on fossil fuels for energy supply, and inefficient water, waste and transportation management accompanied by rapid population growth, often unplanned, have brought increased air and noise pollution, a depletion of water resources and extreme levels of habitat and biodiversity loss in cities. At a much broader level, climate change is a considerable threat for most cities globally, since it tends to exacerbate heatwaves, extreme rainfall events and cyclones, as well as present more dangerous fire risks on the fringes of city environments. Moreover, cities are vulnerable to severe storm surges associated with rises in sea levels and extreme weather conditions. The impacts of such change on the built environment and major infrastructure networks, like transport and energy, could have immediate and damaging effects on urban communities, the urban environment and a city's productivity (Norman 2016).

It should be pointed out that the magnitude and complexity of the problems we face today are far greater than those experienced by the cities of Ur and Shang. In addition to the concrete issues facing city administrators, urban planners and sustainability professionals at the local level, the transgression of boundaries at the planetary scale was not an experience encountered by earlier civilisations in the Holocene. Experts in ecological footprints suggest that our activities can no longer be sustained by one Earth alone. It is estimated that man is now consuming the ecological resources of 1.5 Earths and is heading rapidly towards two Earths (Global Footprint Network 2013). Rockström et al. (2009) believe that three of nine interlinked planetary boundaries have already been overstepped and 'pressures on the Earth System have reached a scale where anthropogenic activities could inadvertently drive the Earth system to a much less hospitable state' (Steffen et al. 2015, p. 2). Therefore, it is important to highlight the importance of earth system and eco-systems in the context of what we argue as the eco-development approach.

Returning to the question asked earlier—Have we learned any lessons from the failures of past civilisations? And have we gained anything from the efforts of archaeologists and climatologists to expose the failings of past civilisations, in addition to mere knowledge? The answer to these questions can be discussed from two very different perspectives. On the one hand, we seem to have learnt nothing tangible. Most of the countries in the developed world continue to regard economic growth as the fundamental criterion for measuring success in society and reluctant to take actions to counteract climate change. One recent example is the growing neglect of the politicians in the US, who argue continuously in favour of their country's prosperity through constant—but not necessarily sustainable—economic growth. Thus, where there are conflicts between development projects and environmental goals or social fairness, it is economic interests that will usually prevail. Across the globe this has been the known 'business-as-usual' mode for many decades. The so-called lessons learnt from the recent projects of the developed world seem to be less applicable for those nations who also aim to have a growing economy. Hence, we witnessed an obvious failure in the Copenhagen talks in 2009. However, on the other hand, there is still some considerable optimism that we be able to avoid the follies of the Ur and Shang civilisations, several millennia after their abrupt demise. Perhaps one key to our 'redemption' is the depth of awareness we have of the issues facing the society in both developed and emerging nations and the increased determination to change the way in which we construct our cities, communities and buildings (Deng et al. 2017). From this perspective, we can throw some light on the future of eco-development and its role in changing the business-as-usual scenarios.

1.2 A Glimpse of the Book

Cities are human settlements that are distinguished by their size, density and heterogeneity, and by the social, political, economic and cultural effects of these qualities (Beall and Fox 2009, p. 32). Archaeologists place the birth of cities, more than 6000 years ago, in the Sumerian region of Mesopotamia (present-day Iraq) on the plains that lie between the Tigris and Euphrates rivers. This region is one where fertile soil and access to waterways from irrigation and transport facilitated surplus agricultural production (ibid., p. 38). Similar to the experience of Mesopotamian cities, urban centres in China served important spiritual, economic and political-administrative functions. At its peak, Chang'an (capital of the Sui and Tanng dynasties, 580–907 AD) may have been home to around one million inhabitants. It was formally planned and carefully regulated its morphology and regimented daily life, reflecting the rigid political hierarchy of the time. As the vast Chinese empire was consolidated through tumultuous episodes of expansion and extraction, cities served critical functions as hubs of culture, intellects, economic, and politics; the functional characteristics that we have seen in many cities of the past and current.

Dating back over two thousand years, Kaogong ji (考工记, The Records of Examination of Craftsman) is a book from the pre-Qin dynasty (before 221 BC) which was used to guide the handicraft industries, including the examination and assessment of craftsmen. It provided a prototype for capital cities:

> *The craftsmen would build a capital city, in a square. Each side was nine li long (about 4.5 kilometres), and accommodated three city gates. The city would be crossed by nine roads perpendicular to nine others and each would be wide enough to allow nine wagons to draw abreast (approx. 16.5 metres wide). The ancestral shrine would be on the left side of the city and the altars on the right. The audience hall would be in the front, and the marketplace in the rear. There would be a quadrangle one hundred paces long between the marketplace and the audience hall.* (Kaogong ji 2012, pp. 112–117)

From the above we can see the functionalities provided by a city were recognised through spatial configurations—the locations of the city gates, residential areas, ancestral shrine, altars, audience hall and marketplace—all linked by an interwoven road network and clearly defined spacing between buildings; hence, a city was seen as spatially based. Visually, there are different spatial scales within a city. These physical boundaries may be visualised as a building site, a building block, a neighbourhood, a district or even a whole city. Buildings, together with support facilities and spaces, are developed to form neighbourhoods that are organised within cities (e.g., regulated by an urban masterplan).

Cities today consume 75 percent of natural resources, produce more than 50 percent of the global waste and account for 60–80 percent of both energy use and associated fossil fuel carbon emissions, while only covering around 3 percent of the earth's land area (UNEP-DTIE 2013). Buildings alone are responsible for more than 40 percent of global energy used, and as much as one-third of global greenhouse gas emissions (GHGs) (UNEP-SBCI 2009). Today, despite their inherent advantages, urban areas are more unequal than rural areas and hundreds of millions of the world's urban poor live in substandard conditions. One particularly striking example is Rocinha (Fig. 1.1), the largest favela in Rio de Janeiro, Brazil which has developed from a shantytown into an 'urbanised slum'. Estimates of its population vary widely, but a median is 225,000 residing in an area measuring 2.23 square kilometres, translating to a density of about one thousand people per hectare (UN Habitat 2011).

We know little about what makes a city sustainable. There is no consensus on how to define sustainability. Nor is there a consensus on what city size, form and spatial distribution of activities best facilitate the rational allocation of natural resources and minimise environment impacts (Alberti 1996; Baeumler et al. 2012). However, there are practices that appear to be prevailing over the 'business-as-usual' cases. For example, according to the World Resource Institute (2014) report, Atlanta and Barcelona, two cities with roughly the same population and economic size, offer two very different types of urban environment: Atlanta's carbon dioxide emissions from private and public transport are 7.5 tons per person, whereas Barcelona's are only 0.7 tons per person. This is largely attributed to the latter city having adopted an urban development model of spatially compact, medium-to-high-density urban form tightly linked by mass transit systems (Green and Stern 2015). This analysis shows the tangible impact of spatial structuring on city-level performance. In addition, it is important to understand and evaluate the interplay between different spatial levels, and as such we can argue that policy development at the city level (and even at the national level) can adversely impact on people's behaviour and change of lifestyles, and ultimately contributing to the enhancement of well-being of the society and the quality of life of its citizens.

Fig. 1.1 Rocinha, the largest favela in Rio de Janeiro, Brazil. Source: UN-DESA (2014, p. 32) World Urbanisation Prospects

China is undergoing the largest scale of urbanisation in history and at an unprecedented pace, particularly from the perspective of intensity of migration from the rural to the urban. Between 1991 and 2012, China's urban population has increased from 26.4% to 52.6%. The urban built areas have expanded from 12,856 to 45,566 square kilometres over the same period, an increase of more than 3.5 times greater in about two decades (National Bureau of Statistics 2014). In this period, a tremendous number of enormous new buildings and mega structures in cities have been constructed to accommodate the increased population. Therefore, it is fairly simple to just recognise these resource and environmental challenges related to the construction and operation of the built environment. On the contrary, these challenges are adhered to the significance of a broad concept of developing a 'resource-conserving and environmental friendly society'. In this respect, China has initiated and developed many policies, strategies and pilot projects at both national and local levels, with focuses on three aspects of: (a) low-carbon/green buildings; (b) new urban areas; and (c) retrofits for the existing contexts. A number of pilot eco-development projects, which seek to de-couple economic growth from environmental degradation by promoting innovative urban policy strategies, practices and technologies, have emerged in the past decade or so. China appears in the front line of reshaping the urban environment (Joss et al. 2011) with around 280 Chinese cities declaring an ambition to develop as an 'eco-city' or 'low-carbon city' in 2012 (China Society for Urban Studies 2012, p. 10). The number is increasing since then.

In this book, we provide a Chinese and interdisciplinary perspective on the relationship between cities and eco-development. Throughout this book, we refer to the spatial levels with the urban context which have traditionally been referred to as separate 'domains'. We recognise the difficulties by using a collective view on them as in current planning they are presented separately. However, the common ground in all of them is to address three key points: of building, space and people, although the distinctions between these may vary across spatial levels.

The following section offers a brief description of the book's structure and the details of individual chapters:

Chapter 2—In the next chapter, we focus on the issues in relation to the themes of sustainability and development, setting the scene for later exploration of key topics, such as the sustainable city, the eco-city and eco-development. In this chapter, we also discuss the topic of eco-development based on its general challenges, implications and actor constellations. The three spatial levels of the built environment are introduced and discussed further, which are then followed up by detailed discussions on the topic of urban sustainability.

Chapter 3—This chapter is dedicated to the concept of eco-development in the global context. We start with some of the globally known examples of eco-development at the three spatial levels, and then explore some of the main methods and globally used evaluation tools for green development. Each of these tools are discussed at their specific spatial level, with further discussion on how they play their role in the promotion of the concept of eco-development at various scales.

Chapter 4—Following on from the previous chapter, this chapter focuses on the context of China. We start with a detailed analysis of China's urbanisation era through a discussion of some of the past and current challenges of urban growth and urbanisation. We then move into the narrative for evolution of eco-development in China, and further discuss the positioning of China in relation to its eco-development strategies. We explore some of the key national policies in relation to China's eco-development, before concluding with one local example.

Chapters 5, 6, and 7—These three chapters serve as our case study chapters at the three spatial levels: macro (city scale), meso (neighbourhood/community scale), and micro (building scale). In each case study chapter, we include several examples of Chinese eco-development case studies, based on a defined categorisation at their spatial level. In Chap. 5, we explore eight cases of eco-city development in China. In Chap. 6, we discuss six cases of Chinese neighbourhood and community projects that have a strong eco-/green agenda. Finally, in Chap. 7, we highlight seven major cases of eco-/green building projects across the country, all of which are well-known and important in terms of their contribution to China's green experimental projects. For all three spatial levels, each case study includes a project description, analysis and project reflection.

Chapters 8 and 9—Chapter 8 serves as a comprehensive reflection to what has been discussed in the previous chapters. Although it focuses mainly on the context and cases of China, it also reflects on some of the major global trends and opportunities. In this chapter, we start by our analytical views at each of the three spatial levels, and by looking at the past and current trends of eco-development at each of them. We then give an overview of benefits from the concept of eco- and/or green from a multiscalar perspective, and particularly discussing the context of China and its past, current and future directions. Later on, we move on to describing and discussing several future projections and potential prognoses which highlight possible directions for China's future eco-development strategies and policy development. Chapter 9 is a offers some brief discussion of some of the paradigms in more detail, along with some brief concluding remarks.

References

Alberti, M. (1996) Measuring urban sustainability, Environmental Impact Assessment Review 1996:16:381–424.
Beall, J. and Fox, S. (2009) Cities and Development. Routledge, London and New York.

Baeumler, A., Ijjasz-Vasquez, E., and Mehndiratta, S. (2012) *Sustainable Low-Carbon City Development in China*. Directions in development; countries and regions. World Bank, Washington, DC. World Bank. https://openknowledge. worldbank.org/handle/10986/12330 License: CC BY 3.0 IGO.

Deng, W., Blair, J., and Yenneti, K. (2017) Contemporary urbanization: challenges, future trends and measuring progress, in Encyclopaedia of Sustainable Technologies, edited by Abraham M. et al. Elsevier Oxford UK.

China Society for Urban Studies (2012) China Low Carbon and Ecological Cities Annual Report. China Building Industry Press, Beijing.

Diamond, J. (2011) Collapse: how societies choose to fail or succeed. Revised edition, New York, Viking Press.

Global Footprint Network (2013) "Earth Overshoot Day". Retrieved March 16th, 2015. Available at: http://www.footprintnetwork.org/en/index.php/GFN/page/earth_overshoot_day/.

Green, F. and Stern, N. (2015) China's "New Normal': structural change, better growth, and peak emissions, the Grantham Research Centre on Climate Change and the Environment and the Centre for Climate Change Economics and Policy, UK.

Joss, S., Tomozeiu, D., and Cowley, R. (2011) Eco-Cities: A Global Survey, Eco-City Profiles, London: University of Westminster.

Kaogong ji (2012) 考工记, The Records of Examination of Craftsman, Dolphin Books, 2012.

Keightley, D.N. (1999) The Shang: China's First Historical Dynasty, In Loewe, Michael; Shaughnessy, Edward. The Cambridge History of Ancient China. Cambridge: Cambridge University Press. pp. 232–291.

Lilienfeld, R. and Rathje, W. (1998) Use less stuff: Environmental solutions for who we really are. New York: Random House.

Marks, R.B. (2012) China: its environment and history. Rowman & Littlefield Publisher UK.

National Bureau of Statistics (2014) China Statistical Yearbook 2014, China Statistics Press Beijing.

Norman, B. (2016) Climate Ready Cities. Policy Information Brief 2, National Climate Change Adaptation Research Facility, Gold Coast.

Rockström, J., W. Steffen, K. Noone, Å. Persson, F. S. Chapin, III, E. Lambin, T. M. Lenton, M. Scheffer, C. Folke, H. Schellnhuber, B. Nykvist, C. A. De Wit, T. Hughes, S. van der Leeuw, H. Rodhe, S. Sörlin, P. K. Snyder, R. Costanza, U. Svedin, M. Falkenmark, L. Karlberg, R. W. Corell, V. J. Fabry, J. Hansen, B. Walker, D. Liverman, K. Richardson, P. Crutzen, and J. Foley. (2009) Planetary boundaries:exploring the safe operating space for humanity. Ecology and Society 14(2): 32. [online] URL: http://www.ecologyandsociety.org/vol14/iss2/art32/

Rosenburg, M. (2016) Largest Cities throughout History. Available at: http://geography.about.com/od/culturalgeography/fl/Largest-Cities-Throughout-History.htm (Accessed 3 December 2016).

Steffen, W., Richardson, K. and Rockstrom, J. (2015) Planetary boundaries: Guiding human development on a changing planet. Science, Vol. 347, Issue 6223.

UN-DESA (2014) World Urbanization Prospects, in: UN (Ed.). United Nations, Department of Economic and Social Affairs.

UNEP-DTIE (2013) Shifting to Resource Efficient Cities: 8 Key Messages for Policy Makers, Cities and Buildings. United Nations Environmental Program – Sustainable Consumption and Production Branch.

UNEP-SBCI (2009) Buildings and Climate Change – Summary for Decision-Makers, in: UNEP-DTIE (Ed.), Paris.

UN-Habitat (2011) Hot Cities: Battle-ground for Climate Change, in: UN (Ed.), Cities and Climate Change – Global Report on Human Settlements. United Nations Human Settlements Programme, Nairobi.

United Nations, Department of Economic and Social Affairs, Population Division (UN DESA) (2015) World Population Prospects: The 2015 Revision, Key Findings and Advance Tables. Working Paper No. ESA/P/WP.241.

World Resources Institute (WRI) (2014) Better Growth Better Climate: The New Climate Economy Report. Washington, D.C.

CHAPTER 2

Sustainability and Development: Challenges, Implications and Actor Constellations

2.1 Introduction

The term sustainability or sustainable development (SD) is relatively new though it is alleged by some scholars that it dates back to some ancient philosophies, such as Taoism in China as well as to the Enlightenment and the Age of Reason. Together with Confucianism and Buddhism, Taoism is one of the pillars substantiating the Chinese traditional culture. Taoist philosophy emphasises respect for nature and the promotion of a harmonious relationship between humanity and nature. The two essential laws of Tao are man is an integral part of nature and man must follow the natural rules to achieve a harmonious state (天人合一, 道法自然). Examples of the latter include the notable book, *Utopia*, written by Thomas More in 1516, and the famous *Essay on Population*, written by Malthus in 1798. However, in practice, this term was hardly heard until the late 1980s, twenty years after the outset of the contemporary environmental movement (Dresner 2002, p. 1). Influential books in this period like Rachel Carson's *Silent Spring*, Paul Ehrlich's *Population Bomb* and the Club of Rome's *Limits to Growth*, have contributed to the discourse that the imbalance between natural resources and greed of modern consumerism is not sustainable. The focus of the term 'sustainability' is derived from theory building and relatively limited practice around natural resource management and alleviation of pollution. It is now expanded to the application of sustainability thinking and methods to the wider problems of integrated environmental, economic and social development.

In 1987, the World Commission on Environment and Development (WCED) and its influential report *Our Common Future* (the so-called Brundtland Report, named after the former Norwegian Prime Minister Gro Harlem Brundtland who chaired the Commission) represents a milestone for the concept of sustainable development on the international stage. This report has hitherto and contributed to the implementation of sustainability at national, sub-national and local levels.

By criticising current development trends that 'leave increasing numbers of people poor and vulnerable, while at the same time degrading the environment' (WCED 1987, p. 5), the Brundtland Report calls for a new path of 'sustainable development'. In the report, the definition of sustainable development comes out as the following passage:

> *Sustainable development is development that meets the needs of the present without compromising the ability of future generations to meet their own needs. The concept of sustainable development does imply limits—not absolute limits but limitations imposed by the present state of technology and social organisation on environmental resources and by the ability of the biosphere to absorb the effects of human activities.* (WCED 1987, p. 8)

This statement provides little information on operationalisation in the practice of sustainable development. However, we can take a distinct impression that sustainability tries to draw a balance between economic development, ecological conservation and social equity, which actually constitute the three pillars of sustainable development (Fig. 2.1). Clearly, sustainable development covers much more than solely environmental protection.

Cities are the key for global sustainability endeavours as they are the largest consumer of resources and contribute the largest proportion of the world's total greenhouse gases (GHG) emissions. A city is an area in which built environment predominates over the natural environment. Indeed, cities are a concentration of buildings where intensive social and economic activities take place, accompanied by a large amount of energy and resource as input which consequently results to waste and pollution as output. Chapter 9 of the Brundtland Report, titled *Urban Challenge*, also states that 'settlement—the urban network of cities, towns, and villages—encompass all aspects of the environment within which economic and social interaction take place' (WCED 1987, p. 243). Therefore, we can argue that cities are physically complex systems in which people, buildings, facilities, hinterlands and various natural and semi-natural environments interact.

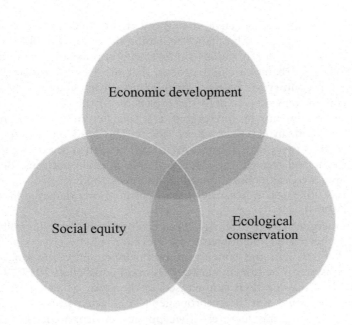

Fig. 2.1 Three sustainability pillars highlighted in the Brundtland Report (drawn by the authors)

Historically, cities as agglomerations of population have gradually evolved to meet the demand for a much broader socio-cultural existence compared to rural environments. Due to their natural origin, traditional cities are more emphasised on organic relationships between people and the environment. However, modern cities, which emerged in line with the industrial revolution, have gained momentum from technological advancement that enables cities to accommodate more people by providing infrastructure and facilities. The geographical size of a city and the effect of population aggregation have been greatly accelerated since then, and now this phenomenon is called 'urbanisation'. In the seminal book, *The City in History*, Lewis Mumford (1961) harshly criticised the current city development model which has caused urban sprawl and other related socio-economic problems. In particular, the form of cities was not organically integrated to the natural environment and to the spiritual values of human community. An ideal city, Mumford argues, should be organically organised and driven by achieving a balance with nature rather than technological innovation.

From a historical perspective, the exponential growth of cities over the past few centuries is a result of cities becoming the destination for fulfilling people's aspiration for material comfort, safety, prosperity and various opportunities. Urbanisation is a process that involves the movement of people from countryside to cities or migration from small towns to big cities. This movement results in the expansion of the urban population and scale. It can thus be interpreted as a social-cultural transition to a higher level human existence (Rees 1999). This kind of transition can further lead to two essential transformations, first, the change of production patterns and second, the change of consumption patterns. Both of them imply increasing consumption of resources and energy because 'the sub-cultures of cities, in many cases, are consumption orientated life style' (Alberti 1996).

The emergence of an explicitly urban ecology is the logical corollary of the two interrelated phenomena: the elevation of the city as principal human habitat, and the concurrent domination and alteration of the earth's ecosystems by human agency (Osmond and Pelleri 2017). Urban sustainability may be then summarised as the study of the interactions between the nature and the anthropogenic components such as building stocks and urban infrastructures—their physical environment as mediated by urban form, together with the tangible and intangible systems—social, economic, technological—which characterise our city habitats (Osmond and Pelleri 2017).

Globally, we can trace confusions between the term 'eco-city' and other terms such as 'green city', 'low-impact city', 'low-carbon city' and 'sustainable city'. While eco-city purely signifies the role of ecology and nature in city environments, through preservation, enhancement, improvement and additions, the other (similar) initiatives do not necessarily focus on the ecological aspect of city planning. Yet we can trace many traditional settlements around the globe that focus on aspects of ecological planning and natural preservation and/or enhancement. As a result, eco-city, as a term, is probably fairly new, but as a concept, it is embedded in some of the very early human settlements.

These terms are used interchangeably in this book to describe the process of eco-development in the urban context. Eco-development is a broader and inclusive term referring to endeavours being taken in the urban context with a focus on three spatial levels—building, neighbourhood/community and city. Fundamentally, all these concepts seek to coherently order different natural and anthropogenic components for various activities in cities to create environmentally responsive, economically feasible and socially inclusive places for communities (Osmond and Pelleri 2017).

We began this chapter with an examination of the origins of urban sustainability. Next we elaborate the dimensions and spatial levels of urban sustainability, followed by a discourse on various actor constellations over sustainability objectives in the urban context. Several global examples at different spatial levels are used to present the current practical effort worldwide. We then turn our attention to the Chinese context by reviewing the current urbanisation trend, and relevant policies at the national and local levels in the next chapter.

2.2 Urban Sustainability in Context

2.2.1 Setting the Scene

When we started talking about the concept of 'sustainable city' in the late 1980s, after it was first coined by Richard Register (1987), the idea was to consider building cities for a healthy future. The emergence of three sustainability dimensions of 'environmental, 'economic' and 'social' came in to consideration soon after. After the Brundtland Report, the concept of sustainable development was acknowledged in discussions of the 1992 Earth Summit in Rio de Janeiro, which led towards the establishment of the eminent international programme of Agenda 21. Divided into four equally important sections, Agenda 21 proposed for: (1) social and economic dimensions, particularly in the context of developing countries; (2) the conservation and management of resources for development as well as the preservation of ecosystems and biodiversity; (3) strengthening the role of major groups; and (4) means of implementation, including several mechanisms for sustainable progress and development. Agenda 21 is a further step to explain the concepts of sustainable development, proposed by the Brundtland Report, at implementation level. This international programme then stimulated the start of major initiatives for city development, including, but not limited to, the ongoing concepts of 'eco-', 'green-', 'resilient-', 'low carbon-', and the currently popular 'smart-'. The term 'sustainable city', however, remains central to all these initiatives; however, it is partially politically controversial in the global arena and partially unfeasible in many contexts. Nevertheless, there are major international programmes, such as the recent 'Sustainable Cities Programme (SCP)', which is a joint UN-Habitat/United Nations Environment Programme (UNEP) capacity-building and institutional strengthening facility. This programme was established for the purpose of supporting local governments, the adaptation of environmental planning management (EPM) and the integration of best practices into [local] legal frameworks and national policies. This programme currently supports more than 66 cities in 10 Asian countries. These are categorised as either SCP demonstration cities or potential replication cases. This and many other similar initiatives still apply Agenda 21 principles, whilst applying multilateral environmental agreements, conventions on climate change and low-carbon transitions at both local and national levels.

Over the course of the past three decades, many researchers questioned the oxymoronic characteristics of the sustainable city. It is, therefore, uncertain that to what extent cities can actually be sustainable; indeed, if they can be sustainable at all. While cities globally generate more than 80% of the gross domestic production (GDP), it is inevitable to see the continuing progress of worldwide urbanisation and city expansions. More importantly, cities are major financial and economic hubs, and, therefore, city development is considered as part of a progressive economic development pattern in the contexts where urbanisation rate is low or not at its peak. Besides, we cannot neglect the fact that we do not have any non-urbanised developed country. As a result, we anticipate further increases in size and number of cities as well as the growing urban population. These will continue to have significant impacts on higher energy consumption, further waste and pollution production, larger resource use, and exacerbated social pressures.

Repeatedly, the city can be seen as an ecosystem with inputs of non-renewable Earth's resources like petroleum, coal, wood, and outputs in the form of garbage, effluent, smoke, gases and heat, among others. Industrialisation, urban growth, and higher income levels means more linear input–output processes, i.e., use of more ecological resources in production and consumption process that ultimately result in waste, reduction of biodiversity and, all types of environmental pollution and resource degradation (Deng et al. 2017). Quantifying the input–output processes in cities is complicated and currently the best way to denote it in a proper scale through the notion of 'eco-footprint' that was firstly developed by Rees and Wackernagel in 1996 in Canada. Foot-printing basically is a quantitative measurement of natural resources and it is used to assess the extent of the impact of human activities on global sustainability.

The global eco-footprint (EF) analysis is an assessment to quantify the ecosystem areas required to support specified human populations. EF quantifies the amount of land area required to sustain the lifestyle of a population of any size, from an individual, household, community, city, country or the world. For example, if we divide the amount of productive land available on the planet (approximately 13.6 billion hectares) by the world population then we would have 2.1 hectares each to provide us with all our resources and absorb all our waste (WWF 2008). Therefore, if an individual is to be ecologically sustainable then their ecological footprint would have to be 2.1 hectares or less. This number can be seen as a target for global ecological sustainability. According to the *Living Planet Report 2016*, the global average ecological footprint is approximately 2.8 global hectares (gha) per capita (WWF 2016) which shows there is an ecological deficiency of 0.7 gha on average globally in 2016. Notably cities usually have a much higher eco-footprint value, particularly in the developed world, for example, 9.8 gha for Calgary, Canada; 7.1 gha for San Francisco, USA; 5.48 gha for London UK, 3.8 gha for Shanghai China (Deng et al. 2017). The reason is because cities cannot be sustained on themselves and demand food, water and other resources from their hinterlands.

Cities today are executing different approaches to reduce their eco-footprints (moving towards sustainability in general) on different scales at local levels. One example is the re-introduction of urban farming. Let us look back first to the two ancient civilisations we discussed in Chap. 1, Ur and Anyang, both failed due to the excessive demand of agricultural yield to feed its growing population. Since the industrial revolution, farming has been gradually driven out of urban areas. Cities today are relying on food import from their hinterlands that increase their footprints. However, in recent years, urban farming has quickly gained prominence in many parts of the world—in both developing and developed countries. Urban farming is a way to reduce the eco-footprint of a city contributing to improve the overall sustainability of a city, while simultaneously improving its agricultural yield. A fully sustained city, however, is still far from the reality. Such urban sustainability practices globally, supported by proper measurement of performance, will probably prevent us from repeating the failure of early civilisations.

2.2.2 Challenges of the Sustainable City

In theory, a city can never be sustainable within its geographical territory as it needs to extract resources from the hinterlands and export its wastes there. As the world continues to urbanise, more resources are required to feed cities, and largely due to inadequate coordination between cities and their hinterlands, environmental problems and sustainable development challenges have been increasingly concentrated in cities, particularly in the lower- and middle-income countries where infrastructure is underdeveloped and sustainability policies have not been implemented. For example, there are many factors that have brought increased air and noise pollution, water resources depletion and habitat/biodiversity loss in cities. Some of these factors can be referred to as inefficient building stock, high dependency on fossil fuels for energy supply, and inefficient water, waste and transportation management accompanied by rapidly growing—and often-unplanned—population.

The increasing ambient air pollutant levels in cities around the world perceptibly affect visibility and productivity. For example in the first week of December 2016, for three consecutive days, air pollution blanketed Paris, leaving its most iconic symbol, the Eiffel Tower, barely visible through smog so thick that the local authorities called it the worst air pollution in at least a decade. Readings of particulate matter (PM10) exceeded 80 micrograms per cubic metre (ug/m^3) while the European Union standard is a maximum daily average of 50 ug/m^3. Another example is China. Increasing traffic congestion, energy consumption and pollution levels in the urban areas are becoming matters of public concern. Only 8 out of 74 major Chinese cities satisfied the national air quality standards in 2014 (MEP 2015). High exposure to unhealthy air can bring morbidity problems like asthma, lung cancer, cardiovascular disease, respiratory diseases, birth defects and premature death. Globally, three million deaths were attributable to air pollution in cities in 2012, about 80–90 percent of these occurring in low- and middle-income countries (Baklanov et al. 2016).

Climate change is a considerable threat for most cities globally, since it causes more heatwaves, extreme rainfall events and intense cyclones, more dangerous fire potential on peri-urban fringes of cities and more severe storm surges associated with sea level rise. The impacts of such change on the built environment and major infrastructure networks like transport and energy could have immediate and damaging effects on urban communities, the urban environment and a city's productivity (Norman 2016). As a consequence of the threat of climate change, cities need to change their planning, design, construction and operation. Argued by Pyke et al. (2007), the traditional planning and design approach, which is typically based on the assumption that climate is static, should change because:

- Climate is changing—trends are toward warmer temperatures, more frequent heat waves, more intense precipitation events, and longer, potentially more severe droughts, and,
- These changes have significant consequences for the performance of the built environment—designs based only on past conditions will encounter significantly different conditions during their expected service lifetimes.

Without coordination between cities and their hinterlands, the regional planning approach may become weak. This eventually puts pressure on potential collaboration opportunities between administrators, urban planners, sustainability professionals and the entire development community at the local scale. This would then lead into barriers of change and an innovative approach to providing critical services like water, sanitation, energy and transportation, and guaranteeing equal access to services for all income groups. And through this, sustainability will not ensue. Moreover, it is evident that also many cities of the developed countries struggle with conventional forms of waste, wastewater and storm water management while providing energy across conventional grids, chiefly fuelled by coal and oil, is reliable but costly and not environmentally friendly. Therefore, these are commonly major issues and bottlenecks of sustainable development in cities.

2.2.3 Eco-City: A Newly Emerged Global Phenomenon

Despite the sustainable city's comprehensive nature and its extended consideration for environmental and ecological protection, its overall framework encompasses a substantial scope for existing cities and city development. The concept of sustainable city may have initiated many sub-initiatives from various perspectives. Its broad framework has provided the platform to explore issues of ecological-friendly development (for eco-city), low-impact development and green economy (for green city), adaptive capacity (for resilient city), carbon management and carbon reduction (for low-carbon city), enhanced performance and quality (for smart city), and many more. All these sub-initiatives have led to substantial action plans and regulatory strategies for new low-impact development, city improvements and retrofits. The presence of sustainability frameworks in such plans and strategies indicate pragmatic concerns of integrated planning that include environmental issues and schemes for the mitigation of climate change and the reduction of GHG (greenhouse gases) emissions. Nevertheless, in recent years, the city labelling of 'eco-', 'green-', 'resilient-', 'low carbon-', 'smart-' or combinations thereof (such as 'Smart Green Resilient (SGR)', smart-eco, smart-green and so on), has become an inexplicable trend rather than widespread implementation of large-scale change for city growth and development. For instance, many Chinese cities are put forward as pilot cities for multiple initiatives of low-carbon, green, eco- and smart simultaneously. However, none of these cities have the capacity or the right strategies to achieve all these goals concurrently, even in the medium term.

For developing cities as in the context of China, there remain general issues for such city labelling or often city branding. In most cases, the success stories are minimal and/or tangible accomplishments are on a small scale or within the boundary of [attractive] demonstration zones. For city planners and policy makers, the challenges are based on four factors: (1) a lack of explicit vision; (2) minimal implementation and sometimes no implementation; (3) not meeting the action plans or targets; and (4) complication with costs and investment attraction. In contrast with the concept of sustainability, the lack of explicit vision is often derived from short-term planning and does not offer clear pathways for sustainable development. From a practical perspective, the lack of implementation and not meeting the action plans are both derived from ambiguous decisions for city growth and development. The cost factor also affects the direction from expectation(s) to reality; therefore, having significant impact on the quality and performance of action plans and targets. On the other hand, we also have many developing cities with no sustainability frameworks, which are still struggling to battle issues of poverty, economic growth, pollution, environmental degradation and so on; and, therefore, are perceived to be in an urgent need for clear direction rather than labelling. The question then is: how can cities go beyond survival and towards 'enhancement'?—This ultimately is the backbone of sustainability framework for city growth and development.

It is difficult to track down when 'eco-city' started as a concept, but there are evidences of eco-city initiatives throughout the twentieth century. The ideas of eco-city planning are certainly visible in major initiatives that were focused on harmonising the city living environments with nature. Now that the world is increasingly urban for the first time (i.e., over 50% since 2008), the harmonisation between the city and countryside/outside of cities is becoming a more recognised matter. Prior to this steady increase in urbanisation, most cities were often not so large or populated. As urbanisation increases, however, we witness more mega cities in terms of both area and population. The management of cities is no longer seen without the relationships between the city and its region as well as the city and its surrounding environments. Moreover, the second half of the twentieth century has brought us significant increases in production, manufacturing and urbanisation. The latter is unprecedented as the increase has been substantial, specifically in countries of the global south.

Eco-city, as a term, was first seen in Richard Register's book of *Ecocity Berkeley: Building Cities for a Healthy Future* (1987), in which he envisioned cities of the future as living systems that enhance biodiversity, environmentally-sound living environments and a healthy city structure. Since then, the term eco-city is used or mentioned in many major reports (e.g., Agenda 21), international events (EcoCity summits), research, education and academic publications. The term has, therefore, become a global city branding term that is used for various city initiatives of small- to large-scale new city-level projects. Similarly, as described by Wong and Yuen (2011), in their book, *Eco-City Planning*, eco-city is a definite term for 'an ecological approach to urban design, management and towards a new way of lifestyle'. This can be interpreted in various ways if it is to be considered from different sectors. In urbanism, it is seen as an environmentally-friendly approach to urban development; in environmental sciences, it can be regarded as urban ecology; and from a scientific point of view, it can be viewed as ecological progress in urban development. Since the birth of the term, there are similar themes that are used in practice that signify the overarching concept of eco-development. These themes include eco-town, eco-village, eco-community, ecological district, eco-neighbourhood, green building and etc. (Roseland 1997; Wong and Yuen 2011).

The concept of 'eco-city', as a newly emerging phenomenon, has been gradually translated into practical initiatives, particularly since the early 2000s (Joss 2011a). Conventional urban environmental efforts have been focused on individual issues such as urban energy, transportation, land use, waste management, water and urban health. In contrast, the eco-city tries to develop an integrated model for urban development holistically. The primary drivers for heightened activity are rapid urbanisation and related climate change concerns (Joss et al. 2013).

While eco-cities have become something of a global phenomenon, it is in Asia that developments have been particularly notable (Joss et al. 2011). Masdar City in the United Arab Emirates and the Sino-Singaporean

Tianjin Eco-city (SSTEC) in China have attracted international attention as the next generation of city development model. In addition, many existing cities have embarked on concerted urban sustainability action programmes and similarly adopted the eco-city label to promote their efforts (Joss 2011a). A global survey conducted by the University of Westminster in 2011 recognised 174 eco-city projects globally, according to the methodical criteria in that research (Joss et al. 2011). Asia and Australasia together have 69 projects, while Europe has 70. There are 25 in the Americas and only 10 in the Middle East and Africa combined. The countries with the largest concentrations are China with 25 eco-city projects, followed by the USA with 17, and the UK and Japan with 16 each (Joss et al. 2011).

China appears to be on the front line with regard to reshaping the urban environment (Joss et al. 2011). In 2012, there were around 280 Chinese cities that have declared an ambition to develop as an 'eco-city' or 'low-carbon city' (China Society for Urban Studies 2012, p. 10). This indicates that many local governments in China have begun incorporating sustainability concerns in the development to improve industrial structures, building energy codes, public transport, and renewable energy generation. It is still too early to judge the success of the eco-city concept in new urban development areas in China as many examples are still under construction, but they are likely to have a significant influence on the planning, design and operation of cities in the future.

In summary, we aim to go beyond the concept of eco-city as a branding mechanism, and explore the current trends and practices of eco-development in the context of China at multiple spatial levels (covered in Chaps. 5, 6, and 7) before discussing the future directions and paradigms of eco-development in China.

2.3 Dimensions and Spatial Levels of Urban Sustainability

2.3.1 Understanding the Built Environment

The Built Environment (BE), as defined in Encarta dictionary (2010), is human-made buildings and structures as opposed to natural features. This definition is general and simple. Visually, the BE encompasses buildings, spaces and infrastructures that have been altered from the natural environment. One more aspect of the BE we should not overlook is users' behaviours and aspirations. Buildings are not standing alone in a vacuum. They stand in a concrete context with complex interrelations to other components of the anthropic-natural system. Individual components of the BE are defined and shaped by context, and simultaneously they contribute either positively or negatively to the overall quality of environments, both built and natural and to the human–environment relationship (Bartuska 2007). In this way, the definition of BE is linked to multidimensional considerations, i.e., environment, society and economy (one more dimension, *governance*, is often added to the urban context, see discussion in the next section), which are generally accepted as the three essential elements of sustainability. The assessment of the BE should encompass social, economic, and environmental dimensions that emerge simultaneously with the erection of buildings.

In practice, the boundary and scope of the BE, may comprise: (a) the collection of various buildings (e.g., residential, office, commercial); (b) the open space between buildings (e.g., roads, parks, playgrounds); (c) occupants and users who impact or are impacted by the existence of the BE; and (d) various infrastructure and services that support the existence of the BE (e.g., energy, water and transport). So defined, the BE can be examined both hierarchically and systematically. The BE is a component of the total city system and interacts with other component systems such as water systems, agricultural systems, and urban natural systems. At the same time, the BE is comprised of its own subsystems, respectively physical building systems, land use patterns, and socio-economic systems. There are also flows of materials, energy, information and wastes across and within the boundary of the urban BE system, which is necessary to maintain and produce a relatively stable state at any particular point in time (sustainable or unsustainable). If necessary, the BE subsystems can be disaggregated further. For example, a building system is physically comprised of a set of elements such as foundations, structures, windows and finishes. On the other hand, a building is one space within an interconnected social and spatial network formed by the urban context the building sits within. To summarise, BE refers to individual buildings extending in spatial scale from a single-standing site to an area with multiple buildings and open space, accompanied by increasingly intensive socio-economic interaction between users and affiliated facilities and urban support services.

BE today means much more than its original connotation such as protecting us from weather and physical attacks. It has become something that is related to the human mind. The BE has become a place where we can meet our aspirations for opportunity, welfare and cultural entertainment. Homes are places where we can retreat from the outside world. We prefer to live in suburban or outskirt areas, where residents possess similar values and social status. Moreover, there is increasing evidence, which indicates that the physical characteristics in and around a built environment directly impact our health. For example, cardiovascular disease, injuries in the home and mental health, are directly related to the indoor environment, house layout and neighbourhood (Jones et al. 2007). In addition, the poor design and maintenance of buildings can lead to acute respiratory illnesses, allergies and asthma, and the so-called 'sick building syndrome' resulting from mould and moisture problems to various indoor pollutants.

In the vision of urban researchers, the BE means much more than its visible physical appearance. Graham (2004, p. 17) describes what urban researchers are thinking when they look at a building:

> When urban researchers look at a building they see the mines, the minerals and forests from which materials are made, they see the road upon which materials have travelled and the power-plants and refineries that supply the fuel for the journey. They also see the rivers that supply our water and which receive our run-off. Urban researchers see the atmospheric emissions caused by the production of the electricity running our building, and by the burning fuels used to transport people to and from it. Most importantly, they see the demands on nature created by the choices we make and know how to make decisions that are life sustaining.

Repeatedly, the role of the built environment and the construction industry in sustainable development gained global attention due to their significant share in global warming, GHG emissions, energy demand and the depletion of non-renewable resources. Given the long lifetime of buildings, choices made today regarding the construction of built environment will have long-term effects, influencing the overall environmental performance for decades to come (Ye, et al. 2013).

2.3.2 Dimensions of Urban Sustainability

The concept of sustainability in an urban context differs from the general term of sustainability that usually addresses three major dimensions—society, environment and economy (Fig. 2.1). One more dimension, *governance*, is often added to the urban context (Faucheux 1998;

Spangenberg 2002; Cheshmehzangi et al. 2017). Governance is also defined as 'institutional' (Spangenberg 2002; Labuschagne et al. 2005; Dawodu et al. 2017), depending on how it is interpreted in the specific context. Furthermore, the institutional dimension highlights the importance of policy making, policy implementation and institutional structures that are, in most cases, the backbone of providing sustainability directions. The consideration of four dimensions of sustainability rather than the original three, enables for more tangible interrelationships between the dimensions in the urban context. For instance, the overlap or the interrelation between the institutional dimension and other three dimensions is perceptible in most of the urban indicators.

In general, sustainability as an integrated approach, can be recognised as an approach to create physically enhanced and socio-economically viable urban environments (Cheshmehzangi et al. 2010). An integrated approach could, therefore, determine the urban form in many ways; for example, less travel to promote more walking and cycling, creating more denser urban fabric, which could result in more open spaces and providing opportunities to enhance the socio-economic base of a city/development (The English Partnerships 2000). The urban form can also play a significant role in climate change and global warming. Therefore, an integrated approach in urbanism could improve sustainability of our urban built environments. However, in most previous academic works concerning sustainable urbanism, environmental performance has only been discussed within an architectural design context (Ritchie and Thomas 2009; Cheshmehzangi et al. 2017). There is, however, a major demand for multispatial understanding of sustainability where the concept is better understood from the interrelationships between the various levels of the built environment.

The environmental performance in an urban context, such as open spaces and public places, has not been discussed widely (Spagnolo and de Dear 2003; Chuang 2008). Comfortable urban spaces should respond to the local micro-climate, and outdoor open spaces should be well designed to maintain comfort for users in the urban context (Nikolopoulou et al. 2004). Only a pleasing physical environment can invite, encourage and facilitate people's activities in an urban context, and consequently vitalise the local community. A sustainable urban design within an integrated approach could reflect on many factors, such as society, health, identity, cultural and even pollution and energy use.

2.3.3 Spatial Levels of Urban Sustainability

2.3.3.1 Spatial Levels of the Built Environment

The built environment is both spatially and temporally based. Visually, there are different spatial scales of the built environment. These physical boundaries may be visualised as a building site, a building block, a neighbourhood, a district or a whole city. A building, a neighbourhood or a city has their own geographical sizes and is linked to the broad spatial matrix through various transport networks. Geographical characteristics can be observed and measured only after a particular location has been specified and many of them are intrinsically coupled with geography, such as infrastructure, topography, and buildings. Daniell et al. (2004) opined that spatial considerations are a crucial factor for successfully assessing sustainability of housing development, and suggested a number of spatial issues that should be included in the assessment, including location, layout, transport, topography, and land use information.

Furthermore, the built environment should also be addressed through a temporal dimension. The physical elements of the built environment are likely to exist for several decades before they are demolished or redeveloped. Both the built entities and the users will consume resources and energy to sustain their existence throughout its service life, and the social, economic and environmental factors affecting the performance of a particular BE could change over time. Life cycle assessment (LCA) is often used to address such temporal considerations.

The built environment has a spatially hierarchical organisation, from a single whole building, to a neighbourhood, an urban district and then a whole city. Moreover, a building can be further desegregated downwards into its generic building systems such as substructure, superstructure, finishes and service systems. In contrast, cities can be examined in a regional context which comprises a cluster of cities. Sustainability assessment should be able to address the built environment at each individual building, neighbourhood or community and urban district or city level.

2.3.3.2 Building Performance and Technologies

According to the US Environmental Protection Agency, the average American spends 93% of their life indoors, including 87% inside a building and 6% inside an automobile. So it is people's basic need to provide and maintain a comfortable indoor environment. In the first place, whether or not any building consumes energy depends upon the climatic conditions within which it stands. In many regions around the globe and in many times throughout a year we need to consume energy to fine-tune the outdoor hostile climates to avoid discomfort. However, the amount of energy consumption is directly linked to design approaches/technologies used. For example, conventionally the base temperature for heating in the UK climatic context is 15.5°C. If outdoor temperature is lower than 15.5°C, energy needs to be consumed to run the heating systems to provide heating. A conventional building in the UK may need heating for eight months in a year. However as the base temperature is in relation to the construction of the building envelope, it may be only 4°C if a building in the UK is built with the Passive House Standard, which is characterised by highly insulated, highly airtight building envelopes coupled with energy recovery ventilation. Accordingly, the heating period may be reduced to less than one month. Technologies can help to improve building performance greatly.

Technologies that are used to improve environmental performance of buildings are generally classified into three groups: passive, active and hybrid. Passive design takes the advantage of climate and maximises the use of natural sources for heating, cooling, lighting and ventilation to create a comfortable indoor environment. Passive technologies have been long used in human history. One example is the Inuit Igloo, which has a compact form to reduce heat loss, use animal skin and snow cover for insulation, and an underground tunnel for entry. This means that the temperature inside the igloo will maintained around 10°C, even if the external temperature drops to −35°C. The modern passive technologies were introduced in 1963 in a publication titled *Design with Climate—Bioclimatic Approach for Regionalism* (Olgyay 1963). It architecturally describes design as a way of tackling the use of natural resources, their human consumption and human reactions to the environment. Passive measures do not involve mechanical or electrical systems and often involve:

- Orientation and shape: compact building type, east–west axis, northern and southern windows, south overhangs, reducing west facing glass, prevailing wind, etc.;
- Insulation and airtightness: proper insulation level, removing thermal bridging, proper window sizing, etc.;
- Lighting: location of openings, maximum use of natural light, daylight control on parameter windows, etc.;
- Ventilation: the wise use of natural wind;
- Thermal mass: the wise use of material masses to make a building responding properly to the change of external climatic conditions.

Active systems, as opposed to passive measures, make use of mechanical or electrical power to run building installations such as boilers, chillers, mechanical ventilation, electric lighting and so on. Renewable systems such as solar power systems and wind turbines are also active systems as they do not reduce a building's energy demand but supply energy from renewable sources to actively reduce carbon emissions. Hybrid systems use some mechanical energy to enhance the use of natural sources; in other words, they use active systems to assist passive measures, for example, heat recovery ventilation, solar thermal systems, radiant facades, and ground source heat pumps might be included in this group. We need to consume energy to run these building installations to generate more energy, for instance, a unit of energy input may generate four units of energy output by a set of ground source heat pumps. Notably, where it is possible to do so, designers and engineers will aim to maximise the potential of passive measures, before introducing active or hybrid systems. This can reduce capital costs for purchasing and running the active and hybrid systems, and also reduce the energy consumed by a building. British architects Brenda and Robert Vale (1991) propose one of the simplest and most straightforward frameworks for green architecture which contains six general principles:

- A building should be constructed so as to minimise the need for fossil fuels to run it;
- Buildings should be designed to work with climate and natural energy sources;
- A building should be designed so as to minimise the use of new resources and, at the end of its useful life, to form the resources for other architecture;
- A green architecture recognises the importance of all the people involved with it;
- A building will 'touch-this-earth-lightly';
- All the green principles need to be embodied in a holistic approach to the built environment.

So to what extent can we call a building green or ecological? There is no universally accepted definition for a green or eco-building. Retzlaff (2009) defines a green building as a structure or group of structures that is designed to increase the efficiency of resource use, including energy, water, indoor environmental quality, siting, infrastructure, and pollution. There is a large number of building assessment systems in current practice which are used to evaluate how buildings affect the environment. They aim to establish standards for green buildings by evaluating performance against criteria (Retzlaff 2008). A typical green building assessment system is comprised of a checklist of 'green measures', combined with their corresponding weightings. The most significant contribution of these assessment systems is to acknowledge the importance of evaluating whole building environmental performance across a broad range of considerations instead of a single criterion such as energy efficiency.

Basically, all these building assessment systems handle the major environmental issues such as energy, water, materials, waste and indoor environment. This is because these are common challenges that each of these tools needs to deal with. On the other hand, there are always divergences with regard to which aspects of the issues should be examined and how much weighting should be given. For example, regarding to the issue of water, 5 points out of 69 (or 7%) are given in Leadership in Energy and Environmental Design (LEED) from the USA, 6 points out of 107 (or 5%) are given in Building Research Establishment Environmental Assessment Method (BREEAM) from UK and 12 points out of 147 (or 8%) in Green Star from Australia. Similarly, in respect of transport, 4 points (or 5%) are given in LEED, 8 points (or 7%) are given in BREEAM and 14 points (or 10%) of the total in Green Star. The reason behind this is that the systems are developed originally to reflect the local contexts such as climates, typical building types, cultures, etc. In addition they are developed on a subjective basis and inevitably affected by the creators' judgements.

Suggestions about the inadequacies of current building assessment systems—which is limited mostly to concern for environmental issues within the building site boundary—are endemic to the literature. One of the most radical criticisms concerning the current green building initiative is proposed by Tuan-Viet (2008), who argues that the expansion in the application of green building rating systems, while pushing buildings as much as possible to the highest rating level, may ignore some other social and economic issues, and, therefore, may possibly not increase the sustainability outcomes of the urban area as a whole. So we need to take a view of building performance in a broader context—neighbourhoods and cities.

2.3.3.3 Neighbourhood Pattern and Community Cohesion
Neighbourhood is a term that is hard to define, but everyone knows a neighbourhood when they see it. According to the Encarta Dictionary, a neighbourhood is a geographically localised community within a larger

city or suburb. Traditionally a neighbourhood is small enough for the neighbours all to be able to know each other. From this definition, it seems that the identity of a neighbourhood is related to two factors: geographical cohesiveness and a place which has the ability to generate social interaction. In practice, a neighbourhood is a relatively small geographical area (compared to a city or a district) and within it a certain level of social functioning occurs. Neighbourhood can be visualised as a residential compound, a village, a business park or an industrial park. The existence of neighbourhood is dependent on the infrastructure and services provided by its urban matrix. Since neighbourhoods, as part of the built environment, are also spatially and temporally based, they should be examined in both spatial and temporal frames.

It is important to recognise that the discussion generally reflects a concept of open community, its geographical size is loosely defined and it is dependent on users' 'familiarity and special distinction' (Humber and Soomet 2006, p. 713), which 'typically go beyond a household's directly adjacent neighbours' (Saville-Smith et al. 2005, p. 13). Unlike such blurred neighbourhood boundaries, 'gated' neighbourhoods that have been seen around the world, particularly in Chinese cities, present a geographical form well defined by walls and gates. However, they are not necessarily known as a sustainable form of neighbourhood design. On the one hand, they are able to provide a higher level of security and sense of community, on the other hand, gated neighbourhoods may lead to discontinuation of urban traffic systems, social segmentation and waste of resources.

A neighbourhood principally consists of individual buildings constructed for various purposes, i.e., residential, commercial and community buildings. Residential buildings are at the centre of a residential neighbourhood, providing shelter for residents. The existence of other types of buildings and the open space is to support the residential function. The residential building typology within a neighbourhood may be diverse, including, for example, multi-family apartment buildings, detached single-storey bungalows, semi-detached houses and townhouses. The commercial buildings provide various socio-economic services to residents such as banking, dining and shopping. Often, they are embraced in a large commercial building or placed on the bottom level of a residential-commercial mixed building with street frontages. This normally forms a podium of commercial units alongside the neighbourhood/community. The community buildings mainly include neighbourhood communication centres, gyms, schools and kindergartens.

Beside buildings, a neighbourhood also includes the open space between buildings such as streets, walkways, lawns and parking lots. Open space is important for people to socialise, do jogging, or walk their dogs. Neighbourhood analysis needs to address how well buildings and the space around them work collectively. For example, a building that overshadows over the adjacent open spaces and buildings can provide additional benefit in summer for people moving around. This is a useful design strategy in Asian cities as they tend to be more populated and compact. Furthermore, considerations should be given to promote 'circulation economy' when an industrial park is under examination. Wastes from a manufacturing plant may be fed into another plant in the park.

As noted above, neighbourhoods are linked to the broader urban context. In order to support their functioning, the urban context needs to provide them with infrastructure such as electricity and water connection, and public transport. Without them, a neighbourhood cannot exist as a viable urban unit. Cole (2010, p. 279) also argues that considering urban context can promote a number of opportunities including exploiting synergies between buildings and other systems and accounting for the social and economic consequences of buildings.

Alongside the functional aspects of buildings, spaces and urban infrastructure, their design, quality and aesthetics all work together to shape neighbourhood and exert a collective influence over the activities and behaviours that take place there. Neighbourhood form and its features have social and environmental consequences. For instance, neighbourhood layout should consider issues, such as facilitating daily lives, integration with scenic axes and skyline of the surrounding areas, and provision of social facilities. It should be noted that a specific neighbourhood form may have both advantages and disadvantages, for example, the gated neighbourhood form may improve security and increase the community cohesion within the neighbourhood, but also promotes segregation from other neighbourhoods and spaces in the city thus cause traffic congestions in the surrounding area.

Finally, beside these physical components, it is necessary to consider the users of neighbourhoods. Unlike inside of a building, neighbourhood should invite and encourage people to interact in the open space, and change the way of consuming energy and other resources through peer-to-peer learning. Understanding social processes around and within the neighbourhood and involving people in partnerships are critical to moving towards sustainability, especially at the neighbourhood/community level.

The neighbourhood's physical existence should encourage interactions between people, encourage a shift of lifestyle and forge environmental awareness. For example, walkable streets around and within a neighbourhood can reduce the use of cars, and community based environment programs are more effective than those top-down directives.

Based on the above discussion, it can be summarised here that the neighbourhood/community, physically, has four basic components: buildings (residential, commercial, community and industrial), infrastructure (e.g., transport, water, waste and information), spaces (e.g., car parking, parks, playgrounds, roads and walkways), and people and their lifestyles. Behind these physical components, there are other not-so-visible factors that also influence the neighbourhood performance such as urban context and energy and resource consumption patterns. For example, big cities may have city-wide rail connection that is a more reliable and efficient public transport means than public buses. Neighbourhoods sitting in the city centre may have better transport services, but poorer environmental quality. Ideally, an eco-neighbourhood pattern can allow children walk to school, adults bike to work, resources to recycle and reuse, and neighbours have more physical activities and social connections.

2.3.3.4 Cities and Urban Planning

A city may comprise a number of neighbourhoods, which are connected through a complex transport network including roads, train lines, biking lanes and sidewalks. In traditional planning practice, a large portion of urban development investment was directed towards the construction of new roads and transport infrastructure to make a new urban area liveable. While urban sprawl and suburbanisation have become a primary urban development in many countries, such an investment is financially unsustainable and cause social segregation and environmental deterioration due to land losses for constructing roads and transport facilities and heavy use of private vehicles. Some urban researchers, such as Peter Newman and Jeffrey Kenworthy, are advocators for the compact city model, which encourages high-density, mixed-use urban form and good access to urban facilities and services. Though there are still some debates about whether the compact city model can be considered as a perfect sustainable urban form, it has been widely incorporated in to many urban development plans, such as California's local planning systems (Tang and Wei 2013). It is also reflected in some latest urban development practice such as Tianjin Eco-city in China. Beside transport systems, a city needs to provide energy supply, water supply, sewage treatment, wastes disposal infrastructure to maintain the resource input–output cycle of the city. Urban planning and design can offer important solutions to reduce the consumption of energy and resources through appropriate land use policies; for example, the exploitation of renewables such as solar, geothermal, wind and biomass, and the reclamation and reuse of rainwater and greywater. Furthermore, city has three essential components: human component, anthropogenic component—the built environment, and natural component. The natural component, such as water catchments and wetlands, is being affected adversely by the continuous encroachment of the ever-increasing human component and the associated built environment. Urban ecology and restoration is a significant consideration in sustainable urban planning, which not only conserve or preserve the natural component of a city, but also remedy the urban ecological losses caused by human activities and restore urban ecosystems including cleaning up contaminated lands, replanting native vegetation and restore water bodies and wetlands. Lastly, as a consequence of resource depletion and environmental deterioration, especially under the threat of climate change, cities need to change its planning, design, construction and operation.

At the core of urban planning is land use and urban design. Sound urban land use planning and urban design strategies can not only save land from development, but, more importantly, they can provide alternative solutions for urban social, economic and environmental sustainability. For example, they are meant to create an overall urban development model, which is compact, liveable, well connected, and public transport oriented. However, the conventional urban planning focuses on the spatial issues, such as, locations, physical forms, massing and scale of the various components of the built environment. Taking China as an example, the whole urban planning process comprises four steps:

- City masterplan: outline the general land use pattern;
- District plan (only for medium and large cities): based on the city masterplan, further land zoning for a district within in a large/medium city;
- Control plan: determination of development intensity of a district (exclude military land and non-developable land), detailed to basic spatial control unit;
- Detailed construction plan: determination of the spatial configuration of a land plot (roads, building, and green areas).

In general, the conventional urban planning system has put great focus on setting out site-specific development parameters such as density, plot ratios, setback requirements, and etc., in the control plan and detailed construction plan. The current mandatory planning parameters in China used in these plans are limited in scope and may not be able to drive the city moving towards sustainability (Stanley 2008). These mandatory site-specific planning parameters include: land use types; building coverage; building height; plot ratio; green space coverage; vehicular access and egress and parking and other facilities. Stanley further comments that these parameters do not have the adequate breadth and depth inside themselves that make them fully relevant to planning issues of eco-cities and sustainable development. Some common examples can be highlighted as: energy usage reduction; use of renewable energy sources; rain water recycling; storm water management best practices; waste management as well as water treatment and reuse. The contradiction rests on the fact that the planning indicators represent the conventional Chinese planning which focuses on physical planning and spatial elements of the city, but little on the resource input/output systems of the city level (Stanley 2008).

2.3.4 Integrated Thinking of Urban Sustainability

In the following sections, we try to explain the key dimensions/issues that need to be considered to sustain an urban setting in the long term and link to the spatial levels where the issues exert explicitly or implicitly. Table 2.1 indicates a comprehensive matrix to highlight the positioning of sustainability dimensions across the three spatial levels of the built environment.

The city is an organised system of many interacting biophysical and socio-economic components, generally known as sustainability issues. To link the issues to the spatial levels is the consideration that the issues are tightly interrelated with a space and can only be explained within the context of a space. These issues interplay with the spatial levels. One individual issue may be more intensively observed at a spatial level but not at the other levels. Each spatial level provides a new level of information to explain a particular sustainability issue. The interplay between the issues and spaces is shown as below in Table 2.2 and Fig. 2.2.

In Fig. 2.2, we classify and explore the spatial levels of the built environment as well as their associated key indicators that are summarised as issues embedded in each spatial level. There are three common outputs

Table 2.1 A matrix to highlight the positioning of sustainability dimensions across the three spatial levels

Spatial Levels	Micro level (building scale)	Meso level (neighbourhood/community scale)	Macro level (city scale)
Dimensions			
1. Governance	Low	Medium	High
2. Social	Medium	High	Low
3. Environmental	High	High	High
4. Economic	Low	Medium	High

Source: Authors' own

Table 2.2 Breakdown of the key issues at each spatial level of the built environment

Building	Neighbourhood/community	City
• Standards • Technologies • Installations • Users	• Patterns • Public open space • Social Cohesion	• Policy • Infrastructure • Governance • Planning

Source: Authors' own

that are described as impacts, energy flows and resource flows. At the building level, the eco/green building design often relied on 'standards' that are either used as regulations or methods of certification (e.g., Leadership in Energy and Environmental Design (LEED) in the US, and Building Research Establishment (BRE) in the UK as the two most globally renowned examples). In order to make our buildings greener, we tend to use 'technologies' that optimise the building's production and consumption rates (such as energy technologies, renewables, water supply and so on). Such technologies are considered viable for the optimisation of key building systems, such as energy, ventilation, water and lighting. These systems require 'installations' that are vital to any green building design. And, finally, it is important to consider the 'users/end users', often known as building occupiers, whose behavioural pattern and consumption needs and preferences are key to the sustainable operation of a building.

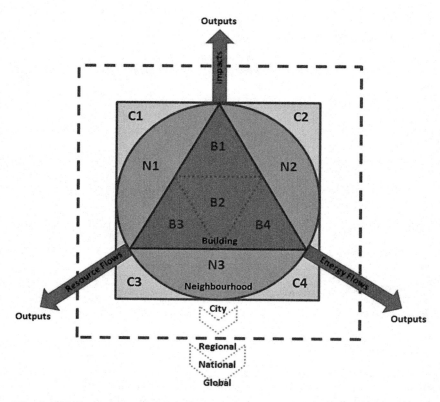

Fig. 2.2 The interplay between the issues and spatial levels of the built environment, at three levels of 1. Building level (B), 2. Neighbourhood level (N), and 3. City Level (C). The issues are divided into four at the building level (namely: B1. Standards, B2. Technologies, B3. Installations, and B4. Users), three at the neighbourhood/community level (namely: N1. Patterns, N2. Public Open Space, and N3. Social Cohesion), and four dimensions at the city level (namely: C1. Policy, C2. Infrastructure, C3. Governance, and C4. Planning). Source: Drawn by the authors

At the neighbourhood/community level, the main focus is on the key aspect of 'patterns' that include neighbourhood forms, neighbourhood layout and related spatial qualities. The other aspect is the overlapping aspect of 'public open space', where the environmental and social values are significant. The last aspect at the neighbourhood level is 'social cohesion', which is a key aspect of mixed-use communities, business districts of mixed nature, and residential districts. At the city level, the power of 'policy' is more tangible than in the other two spatial levels. Similarly, the importance of 'infrastructure' for city structures, 'governance' for city management, and 'planning' for city development are key to eco-/green development at the city level.

From the above discussion, it appears that a sustainable urban development should be able to address the three spatial levels—building, neighbourhood/community and city. Concerted and integrated effort should be made across the three levels. At building level effort is more focused on new construction techniques and building technologies. At neighbourhood and community levels, the design of neighbourhood patterns should be emphasised on key aspects of accessibility, connectivity, and community cohesion. At city level, focuses are given to urban planning policies. Urban sustainability, as a holistic approach addressing the three spatial levels, should be one that is accessible, manageable, environmentally friendly, socially viable and economically efficient. The sustainability factors addressed across the three spatial levels generally include:

- **Providing sound and healthy environmental quality**—This should be provided through effective urban sustainability policies, urban land planning and urban design. Key issues include clear policy direction towards sustainability, implementation of urban sustainability targets, effective urban sustainability governance, optimised urban land planning and efficient public transport.
- **Increasing density in urban areas**—This can protect valuable ecological areas by reducing sprawl, reducing the amount of land that is developed, improving the viability of town centres and public transport and directly affecting travel behaviour;
- **Reducing car dependency as a priority of neighbourhood pattern**—Related issues include mixed-use development, proximity of daily used services and facilities, availability of effective, safe and convenient public transport, neighbourhood walkability, use of bicycles and a reasonable urban road network;
- **Providing communication facilities and quality public spaces**—This factor plays a key role in neighbourhood sustainability. They can encourage people to interact and forge a sense of community, and

improve the satisfaction of residents. Public space also provides local habitat, facilitates the use of rainwater, increases walking and is the stage for creative activities;
- **Providing efficient technological solutions to energy, water and materials in buildings**—Related issues include installing efficient appliances, onsite generation of renewable energy, reuse of water and materials, and enhancement of building insulation.

2.4 Actor Constellations of Sustainability

2.4.1 Urban Sustainability: Actor Constellations

As discussed in the earlier sections, cities are not self-sustained. They rely heavily on their hinterlands to provide resources for sustenance. For example, supplying wood for making furniture in urban households can speed up deforestation. Improving water supply and sanitation in cities can increase the drain on water resources in the region of which it is a part. Nevertheless, solving an environmental problem in cities is not necessarily a step in the right direction from a global perspective (Beall and Fox 2009, p. 165). Such a natural demand has increasingly shaped the environments of the hinterlands, which used to be the regions surrounding a city. Under globalisation and the convenience of massive transportation means, these hinterlands may be thousands of miles far from the cities they provide supplies. Thus, urban sustainability should be viewed from the perspective of people living and working in cities and the regional, national and global context at multiple levels. In this respect, the relationship between them may work in different directions with different sustainability objectives.

Urban institutions can take actions to provide services, conserve water, recycle waste and reduce greenhouse gas emissions. However, they cannot be held responsible for reducing climate risk beyond their jurisdictions. What urban local government can do, is to work on their adaptive capacity, being 'the potential of a system or population to modify its features or behaviour to cope better with existing and anticipated stresses' (Beall and Fox 2009, p. 165). This involves both planned interventions and systems, both reactive and anticipatory, such as the rapid restoration of infrastructure or adapting land-use planning and regulatory frameworks to reduce the vulnerability of urban dwellers. It also means being responsive to the spontaneous adaptations made by individuals and groups within cities (Beall and Fox 2009).

Joss (2011b) examines the multiple actors involved in the development of two large scale eco-city projects in the USA—Sonoma Mountain Village (SOMO) in Sonoma County, California and Treasure Island (TI) in San Francisco, California. Both projects have involved private sectors though the level of involvement is different—TI is formally based on a public–private partnership (PPP) agreement and the other is purely a private development. In the development of the TI project, the Treasure Island Development Authority (TIDA), a governmental agency of the State of California and owner of the land, entered into an exclusive PPP agreement with Treasure Island Community Development (TICD), a private consortium acting as the master developer. On entering the PPP, TICD has a key role in the project development and delivery. Urban sustainability objectives have been centrally integrated in the agreement together with the business goals. The land was conveyed from TIDA to TICD for free. However, TICD is required to pay all upfront costs and various sustainability and public benefit measures, such as affordable housing, the creation of parkland and onsite renewable generation, all public spaces including the affordable homes are still owned by TIDA, which will repay these over three decades through tax increments and service charge income generated from new residents and businesses. TIDA also led extensive public consultation sessions to invite comments from local residents.

SOMO is more private compared to TI. A regional building company is the sole owner of the land, master planner and sustainability innovator. The local government was involved in a mainly regulatory role such as approving the masterplan and zoning plan. The wider state-level actors were more loosely involved in the amendment of the current state energy policy to allow the use of a single solar array for local distribution to homes. The building company also entered a community benefit agreement with the Accountable Development Coalition (ADC), a regional non-governmental organisation, to address environmental and social interests. There no public underwriting for SOMO project. Instead, the project is entirely dependent on revenues from property leasing and selling upon completion, thus it may be affected significantly by market variation.

It should be also noted that both TICD and CE have incorporated sustainability goals into their business models, thereby recognising that urban sustainability is an opportunity rather than a risk. Both projects have also been engaged with international actors. They are not limited to satisfying the US LEED-Neighbourhood Development (LEED-ND) certification, but also go through wider international discourses and processes. SOMO has involved BioRegional, originally from the UK, and is the evaluator of the famous eco-development of BedZED back in the 1990s. BioRegional encouraged CE to consider developing SOMO beyond LEED-ND Platinum and seek endorsement by One Planet Communities, a more stringent certification. TI's international partner is the *Clinton Climate Initiative* which selected TI as one of the 16 founding projects under its Climate Positive Development Program. TI was also certifies as a LEED-ND Gold project.

The delivery of eco-projects needs to assemble around a set of agreed sustainability objectives, targets and this involves multiple actors. Governments used to be at the centre of such initiatives as the incremental cost of developing eco-projects held back private companies. Currently, a green market has emerged in many countries and corporate awareness of sustainability is growing (Global Reporting Initiative 2011). For example, as many as 21 stock exchanges across the world could introduce sustainability reporting standards in the coming months. They would join the 17 exchanges that currently recommend listed companies report on environmental, social and governance (ESG) issues as well as providing model guidance on sustainability to participating companies (Khalamayzer 2016).

On one hand, city authorities are not willing to carry the financial costs and related risks. It is often seen that they are also lack of experience and technical resources to deliver the eco-projects. On the other hand, private companies seize the opportunity to get involved as investors, developers and master planners; as a result of which urban sustainability is substantially incorporated into new business models (Joss 2011b). Joss further comments that the privatised urban sustainability has been observed in many large eco-development initiatives around the world. Nevertheless, the constellation of public and private actors, and collective effort from local, regional, national and international levels, will move our cities towards a unified set of sustainability goals. As stated by Joss (2011b, p. 346):

> The factor that sustainability deals with, and cuts across, the economic, social and environmental pillars of policy making—and does so at multiple level from the local to the global and involving a mixture of state and non-state actors—has prompted calls for more synergistic approaches to developing policies and implementing decisions than is the case of more traditional 'command-and-control' policy and decision making.

Accordingly, it is clear that, without close collaboration from all sectors of society, sustainability is simply impossible to achieve. This is particularly depending on managerial elements of government at federal, state and local level to engage with goals that pursue sustainability, and mobilise non-government organisations, academia, interested parties and individuals in civil society, to work collectively towards environmental responsibility and social equity in an economically effective way.

2.4.2 Eco-Development: Actors and Barriers

The actor constellations of eco-development are drawn from across sectors and disciplines, from policy makers, planners, designers, manufacturers, developers, builders to users, property managers and service providers, and so on—all are part of the course of delivering, engineering, running and maintaining a development. As discussed earlier, the built environment should be also addressed through a temporal dimension. This further requires an integrated whole lifecycle assessment (LCA) approach (Bayer et al. 2010) that accounts for all impacts on environmental protection, social wellbeing and economic prosperity through the following four phases:

1. Integrated planning and design
2. Construction
3. Operation and maintenance
4. Reuse/demolishing

Using the lifecycle framework from Bayer et al. (2010), Deng et al. (2016) enumerated the key stakeholders in each phase and the barriers of developing a green building project from a lifecycle perspective (Fig. 2.3). The planning and design phase often involves two substages: concept definition and design. Key stakeholders include public authorities/green building councils, clients/investors/property developers, and design professionals/green building consultants/specialists. Barriers recognised in the literature (BCA 2010) during this phase include: lack of market recognition, high financial risk, lack of progressive policies and favourable incentives, high technological challenges and lack of qualified professionals, inadequate access to relevant knowledge and technologies, lack of communication and leadership, and high cost of green building products.

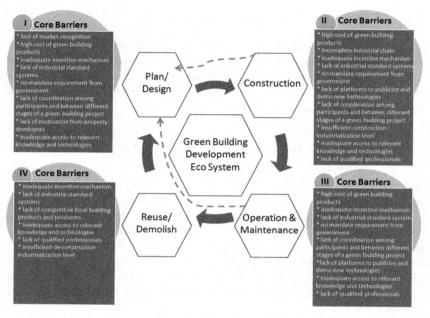

Fig. 2.3 Green building development ecosystem—key barriers through the whole building life cycle. Source: Deng et al. (2016) (drawn by one of the authors)

Different from the conventional construction process, eco-development projects change generally sequential processes to an integrated interactive processes; utilise the skills and knowledge of suppliers and constructors effectively in the design and planning of the projects; facilitate effective decision-making and efficient communication in eco-development project management (CTF 2014). Key stakeholders at the construction phase comprise developers, green building consultants/specialists, contractors and subcontractors, and material and equipment suppliers. The barriers and risks associated to them to take on complex green projects are:

- Insufficient green construction industrialisation level;
- First-mover risk;
- Lack of platforms to publicize and demonstrate new technologies;
- High cost of green building products and systems;

- Traditional linear procurement process;
- Lack of coordination among participants and between different stages; and
- Lack of knowledge and trust between stakeholders.

Opportunities for green interventions could be opened up through public authority-supported market research and the establishment of big data platform for seeking partnerships and public engagement. Based on successful case studies (OECD 2014), alternative procurement models for green building projects would assist design-making on progressive green building policies in the public and private sectors. Education for green living and workforce training, district-scale renewable energy generation and rainwater harvesting would enhance green job market creation and equip the workforce with the necessary technical skills.

All stakeholders in eco-development projects take on the associated risks in time, cost, quality and technical issues, organisation and management, policy and standards, safety, ethics and reputation, and the environment (Yang and Zou 2014), their interaction not only relate to steering policies, technical competence, technology readiness and the modernisation of building industries, but also link to effective risk mitigation actions through the social network.

Eco-development in-use brings the following stakeholders to work together for intended green performance during operation and maintenance phase, including building designers and constructors, building owners, operators/facilities managers, building occupants, and green building systems suppliers and installers. They are all involved in the soft landing process in helping to solve the performance gap between design intentions and operational outcomes. The barriers associated to the building operation phase are:

- Lack of incentives and market recognition;
- Inadequate access to relevant knowledge and technologies;
- Insufficient knowledge and skills training and qualified professional pool on intelligent facility management;
- Lack of platforms to showcase successful technologies and operational templates
- No mandate regulations on commissioning and soft landing, and,
- Lack of social pressure on green living behaviours.

Financial incentives for green leases, green facilities management, benchmarking and follow-up, and green criteria in asset valuation for noticeable market advantage (UNEP 2014) are highlighted as green interventions to promote sustainable operations during a building's lifetime. UK Cabin Office (2013) set up a series of frameworks and guidelines to assist the construction industry and its clients deliver better buildings, and help in bridging the performance gap between design intentions and operational outcomes. It has been revealed that the high initial investment of eco-developments can be 'paid back' during the operation phase through energy saving, higher rents, less maintenance and longer lifespans. Meanwhile they can bring higher comfort and well-being to their occupants.

Several options are available to extend a development's life or ensure it is disposed of safely, including refurbishment, reuse and recycling, and final disposal. Stakeholders involved in this reuse/demolishing phase of a building lifecycle are mainly policy makers, design professionals, construction contractors and material, equipment and system suppliers. Depending on specific construction materials, such as concrete, metal, timber, plastics, glass, etc., end of life recycling and disposal measures can be diverse, involving effective industrial supply chain management. Cradle-to-cradle approach challenges building professionals and demand green thinking in advance. Current barriers to such systems are the lack of industrial standard systems, an inadequate incentive mechanism, the lack of market recognition, inadequate access to relevant knowledge and technologies, and lack of platforms to promote leading industry best practices.

The whole building lifecycle assessment (LCA) approach facilitates effective communication and prompt decision making by sharing stakeholders' experiences, knowledge and expertise. These are recognised as key success of the iterative processes for the best solutions of sustainable development. Design analysis tools, such as, energy performance modelling, natural ventilation strategies, and daylight simulations, among others, enable the optimisation of building performance and the adequate scientific testing of design options. Utilising an integrated design and management approach with robust tools to ensure interactive communication, knowledge and experience sharing between stakeholders will overcome existing barriers in terms of eco-development.

REFERENCES

Alberti, M. (1996) Measuring urban sustainability, Environmental Impact Assessment Review, volume 16, pp. 381–424.

Baklanov, A., Molina, L.T. and Gauss, M. (2016) Megacities, air quality and climate. Atmospheric Environment 126.

Bartuska T. J. (2007) The Built Environment: definition and scope, In: McClure W., Bartuska T. J., Young G.L. (eds.), *The Built Environment: A Collaborative Inquiry into Design and Planning*, Wiley, John & Sons Incorporation.

Bayer, C., Gamble, M., Gentry, R. and Joshi, S. (2010) AIA guide to building Life Cycle Assessment in practice, The American Institute of Architects. Available at: http://www.aia.org/aiaucmp/groups/aia/documents/pdf/aiab082942.pdf.

BCA (2010) Building planning and massing – Green Building Platinum Series, Building and Construction Authority (BCA) Singapore, Available at: www.bca.gov.sg/GreenMark/others/bldgplanningmassing.pdf.

Beall, J. and Fox, S. (2009) Cities and Development, Oxon: Routledge.

Cheshmehzangi, A., Zhu, Y. and Li, B. (2010) *Integrated Urban Design Approach: Sustainability for Urban Design*, in the proceedings of ICRM 2010, the 5th International Conference for Responsive Manufacturing, Ningbo, China, 11–13 January 2010.

Cheshmehzangi, A., Zhu, Y. and Li, B. (2017) Application of Environmental Performance Analysis for Urban Design with Computational Fluid Dynamics (CFD) and Eco Tect Tools: The Case of Cao Fei Dian Eco-City, China, *International Journal of Sustainable Built Environment*, in press.

China Society for Urban Studies (2012) China Low Carbon and Ecological Cities Annual Report. China Building Industry Press, Beijing.

Chuang, C. M. (2008) *The Research of Outdoor Thermal Comfortableness in Summer – Tainan Country, Tainan City and Kaohsiung City Outdoor Spaces as Case Studies*. Master Dissertation of Department of Architecture, NCKU.

Cole, R. J. (2010) Environmental assessment: shifting scales, In Edward Ng (eds) *Designing high-density cities for social and environmental sustainability*, pp. 273–282, Earthscan London.

CTF (2014) Rethinking construction – The report of the Construction Task Force to the Deputy Prime Minister, John Prescott, on the scope for improving the quality and efficiency of UK construction. Available at: http://constructingexcellence.org.uk/wp-content/uploads/2014/10/rethinking_construction_report.pdf.

Daniell K.A., Kingsborough A.B., Malovka D.J., Sommerville H.C., Foley B.A. and Maier H.R. (2004) A review of sustainability assessment for housing development, Research Report No. R175, The University of Adelaide Australia.

Dawodu, A., Akinwolemiwa, B. and Cheshmehzangi, A. (2017) A Conceptual Re-Visualisation of Sustainability Pathways for the Development of Neighbourhood Sustainability Assessment Tools (NSATs), *Journal of Sustainable Cities and Society*, Vol. 28, pp. 398–410.

Deng, W., Blair, J. and Yenneti, K. (2017) Contemporary urbanization: challenges, future trends and measuring progress, in *Encyclopaedia of Sustainable Technologies*, edited by Abraham M. et al. Elsevier, UK.

Deng, W., Yang, T., Llewellyn, T. and Tang, Y. T. (2016) Barriers and policy recommendations for developing green buildings from local government perspective: a case study of Ningbo China, Intelligent Buildings International, pp. 1–17, https://doi.org/10.1080/17508975.2016.1248342.

Dresner, S. (2002) *The principles of sustainability*, Earthscan Publications Ltd, London, UK.

Faucheux, S. (1998) *Intergenerational equity and governance in sustainable development policy*, in Proceedings of the 5th Biennial Meeting, International Society for Ecological Economics, November 15–19 1998, Santiago, Chile.

Global Reporting Initiative (2011) Sustainability reporting guidelines, Version 3.1: Available at: https://www.globalreporting.org/resourcelibrary/G3.1-Guidelines-Incl-Technical-Protocol.pdf (Accessed: 15 December 2016).

Graham P. (2004) *Building ecology: first principles for a sustainable built environment*, Blackwell Publishing, London.

Humber, W. and Soomet, T. (2006) The neighbourhood imperative in the sustainable city; In Mander, U., Brebbia, C. A., and Tiezzi, E., (eds.), The sustainable city: urban regeneration and sustainability, Wit Press, pp. 713–722.

Jones P., Patterson J., and Lannon S. (2007) Modeling the built environment at urban scale: energy and health impacts in relation to housing. Landscape and Urban Planning 83, pp. 39–49.

Joss, S. (2011a) Eco-Cities: The Mainstreaming of Urban Sustainability; Key Characteristics and Driving Factors, International Journal of Sustainable Development and Planning, 6:3 (2010) 268–285.

Joss, S. (2011b) Eco-City Governance: A Case Study of Treasure Island and Sonoma Mountain Village, Journal of Environmental Policy & Planning, 13:4, 331–348, https://doi.org/10.1080/1523908X.2011.611288.

Joss, S., Tomozeiu, D. and Cowley, R. (2011) Eco-Cities: A Global Survey, Eco-City Profiles, London: University of Westminster.

Joss, S., Kargon, R. H., and Molella, A. P. (2013) From The Guest Editors, Journal of Urban Technology, 20:1, 1–5, https://doi.org/10.1080/1063073 2.2012.73540.

Labuschagne, C., Brent, A. C. and van Erck, R. P. G. (2005) Assessing the sustainability performances of industries, *Journal of Cleaner Production*, Volume 13, Issue 4, pp. 373–385.

Lewis, M. (1961) *The City in History, its Origins, its Transformation, and its Prospects*, London: Secker & Warburg.

Nikolopoulou, M., Lykoudis, M. and Kikira, M. (2004) Centre for Renewable Energy Source, Greece.

Norman, B. (2016) Climate Ready Cities. Policy Information Brief 2, National Climate Change Adaptation Research Facility, Gold Coast.

OECD (2014) Compact City Policies: Korea: Towards Sustainable and Inclusive Growth, OECD Green Growth Studies, OECD Publishing.

Olgyay, A. and Olgyay, V. (1963) Design with Climate – Bioclimatic Approach for Regionalism, Princeton University Press.

Osmond, P. and Pelleri, N. (2017) Urban ecology as an interdisciplinary area, in *Encyclopaedia of Sustainable Technologies*, edited by Abraham M. et al. Elsevier, UK.

Pyke, C. R., Johnson, T., Scharfenberg, J. and Groth, P. (2007) Adapting to climate change through neighbourhood design, GTC Energetics Inc.

Rees, W. (1999) The built environment and the ecosphere: a global perspective, Building.

Register, R. (1987) Ecocity Berkeley: Building Cities for a Healthy Future, Berkeley, CA: North Atlantic Books. Research & Information, 27(4/5), pp. 206–220.

Retzlaff R. C. (2008) Green building assessment systems: a framework and comparison for planners, Journal of the American Planning Association, Vol. 74, No. 4, pp. 505–519.

Retzlaff R. C. (2009) The use of LEED in planning and development regulation, Journal of Planning Education and Research, Vol. 29, No. 1, pp. 67–77.

Ritchie, A. and Thomas, R. (2009) *Sustainable Urban Design: An Environmental Approach*, Oxon: Taylor and Francis.

Roseland, M. (1997) Dimensions of the ecocity. Cities, 14(4): 197–202.

Saville-Smith, K., Lietz, K., Bijoux, D. and Howell, M. (2005) Neighbourhood sustainability framework: prototype, NH101, Beacon Pathway Limited.

Spagnolo, J. and de Dear, R. (2003) A Field Study of Thermal Comfort in Outdoor and Semi-Outdoor Environments in Subtropical Sydney Australia, *Building and Environment*, 38, pp. 721–738.

Spangenberg, J. H. (2002) Institutional sustainability indicators: an analysis of the institutions in Agenda 21 and a draft set of indicators for monitoring their effectivity, *Journal of Sustainable Development*, Volume 10, Issue 2, pp. 103–115.

Stanley, C. T. Y. (2008) Planning for Eco-Cities in China: Visions, approaches and challenges, 44th ISOCARP Congress, Dalian China, September 2008; http://www.isocarp.net/Data/case_studies/1162.pdf.

Tang, Z. and Wei, T. (2013) The history and evolution of eco-city and green community, in Tang, Z. (eds), Eco-city and green community: The evolution of planning theory and practice, Nova Science Publishers, New York.

Tuan-Viet D. (2008) Design for sustainable cities: the compact city debate and the role of green building rating systems, EcoCity World Summit 2008 Proceedings.

The English Partnerships (2000) *Urban Design Compendium*, Volume 2. London: Llewelyn-Davies.

The World Commission on Environment and Development (WCED) (1987) Brundtland Report: Our Common Future, Oxford University Press, Australia.

UK Cabinet Office (2013) "Government Soft Landings Section 5 – Environmental Management." http://www.bimtaskgroup.org/wp-content/uploads/2013/05/Government-Soft-Landings-Section-5-Environmental-Management.pdf.

UNEP (2014) Greening the supply chain, United Nations Environment Programme, Sustainable Buildings and Climate Initiative (UNEP-SBCI) – Promoting Policies and Practices for Sustainability. Available at: http://www.unep.org/sbci/pdfs/greening_the_supply_chain_report.pdf.

Vale, B. and Vale, R. (1991) Principles of green architecture, in Wheeler, S. M. and Beatley, T. (eds) (2009) The sustainable urban development reader, Oxon: Routledge.

Wong, T. C. and Yuen, B. (2011) *Eco-City Planning: Policies, Practice and Design*, London and New York: Springer.

World Wildlife Fund (WWF) (2008) Living Planet Report 2008, The World Wildlife Fund.

World Wildlife Fund (WWF) (2016) Living Planet Report 2016, The World Wildlife Fund.

Yang, R.J. and Zou, P.X.W. (2014) Stakeholder-associated risks and their interactions in complex green building projects: A social network model, Building and Environment (73): 208–222.

Ye, L., Z. Cheng, Q. Wang, W. Lin, and F. Ren. 2013. "Overview on Green Building Label in China." Renewable Energy 53: 220–229.

2.4.3 Websites

Encarta dictionary (2010) Available at: http://encarta.msn.com/encnet/features/dictionary.

Khalamayzer, A. (2016) Sustainability reporting comes of age. Available at: https://www.greenbiz.com/article.

MEP (2015) official website of the Ministry of Environmental Protection. Available at: http://www.mep.gov.cn/gkml/hbb/qt/201502/t20150202_295333.htm.

CHAPTER 3

Eco-Development in the Global Context

While the built environment has a considerable influence on environmental resilience, economic affluence and social well-being, global efforts have been taken in order to change the ways of planning and designing our cities, neighbourhoods and buildings. This especially has become of a major concern since the discourse of global warming was promoted in the 1990s. For example, in 1990, the world's first established tool of assessing and certifying the sustainability performance of buildings, Building Research Establishment Environmental Assessment Method (BREEAM), was published by the Building Research Establishment in the UK. In 2002, the first zero-energy neighbourhood development, BedZED, was completed in London. This chapter aims to present the concept of eco-development in the global context, and some globally known examples at the city, neighbourhood and building levels. This chapter will further explore some of the internationally-used tools and approaches for evaluating the sustainability performance of the built environment at the three spatial levels, respectively. Many of these tools and approaches have been developed in recent years and represent the current best practices of achieving higher sustainability performance through planning and design.

3.1 Common Threads from the Global Examples

Globally, there have been some examples and success stories that commonly aim to address the goals of sustainable development. One major movement is that of eco-system planning, which has been embedded in the planning agendas of many city around the globe. Initiated by Richard Register back in 1987, an ecologically-sound city development should be resourceful, possess clear future vision and be composed of natural environments and biodiversity. The role of nature-based approaches in the city environments, which is mainly related to the diversity and biodiversity of cities, is highlighted as a healthy factor in urban planning and development. Thus, a large body of urban land is required for natural species of various types in the city environments which can nurture biodiversity in and around cities. Furthermore, an eco-development is resourceful, protects the existing, and nurtures future opportunities. The vision is, therefore, holistic in terms of how a city may develop or redevelop. As presented by Gibson et al. (1997), in the argument on 'putting cities in their place', such a planning approach focuses on ten principles that address all aspects of sustainable development. Nevertheless, an eco-system-based approach in the built environment highlights the significance of nature in the planning process from an interdisciplinary perspective. Such planning enables eco-friendly community transitions towards the enhancement of institutions, and the development of social learning and social capital, building on the community structure and environmental enlightenment; all of which we consider in the discussed four dimensions of sustainability in the built environment.

For the purposes of discussion, it can be opined that the concept of city is very much linked with the 'urban ecology' movement, an initiative that was started in the 1970s by Richard Register and his team. Through their research, they defined urban ecology based on ten principles which were later adopted as indicators of sustainable development. At first, they argued for the revision of land uses and land use priorities in order to allow more room for compact city environments. By doing so, more land could be utilised for diverse, green and pleasant urban environments, some of which could be mixed-use or transit-oriented developments (TODs). By creating such urban structures, we could then revise our transportation infrastructure and priorities. More space can be allocated to walkable and cycleable environments where new developments could then focus on accessibility and enhanced mobility in the city environments. An ecological city would be healthier, with restored urban environments, particularly those natural environments that can be protected and restored where needed. A particular emphasis is given to urban gardening, urban greening development projects and local agriculture opportunities, with which the benefits are threefold and not only environmental. A mixed-development approach is also encouraged as to promote affordability and convenience for the citizens. As part of this movement, social justice and social improvement are also considered as part of the social dimension of eco-development. The other aspects focus on key urban innovations, such as recycling possibilities, technology adaptation, resource conservation and so on. Some of these are addressed to supporting ecologically-sound economic activities and reducing the levels of both pollution and waste. One major part also focuses on environmental education, and increasing the awareness of the general public in regard to ecological sustainability issues.

Another common approach to eco development is 'green economy' and working for achieving healthy economies and healthy ecosystems (Vodden 1997). For instance, an earlier sustainable economy initiative in London includes key activities that range from green home and business check-ups, to the development of key enterprises and training that address the environmental education. In addition, environmental entrepreneurship is highlighted in several cases of green economic development. In this respect, we can argue that environmental entrepreneurship, as an industry, has blossomed due to the need for such eco-development. These include the monitory of production, consumption analysis, as well as industries for protection and restoration of the natural environments. Also known as 'circular economy', the green economic development is considered to be one of the key pathways to sustainability.

Moreover, the sustainability factors are mainly developed based on current practices and situations. A similar recent approach is the development of Sustainable Development Goals (SDGs), which address the current issues and challenges for a healthier transformation of our societies by 2030. In the case of the built environment, the sustainability factors are also associated with what takes places on the ground. Some of these current practices are discussed in Chap. 2, widely addressing the issues of urban density, transportation, neighbourhood planning, the provision of communication facilities and community spaces, the provision of efficient solutions for our built environments, increased economic efficiency, and the provision of healthy environmental quality in cities.

The following section presents six well-known examples at the three spatial levels, respectively, representing the current global best practices of eco-development. These projects were completed between the late 1990s and the early 2010s, reflecting a key time period in which eco-development moved gradually from theory to practice. They provided exemplary models and principles for developing eco-development in other contexts. For example, BeDZED, Britain's best known eco-neighbourhood, has been exhibited in the 2010 World Expo in Shanghai. China has published its own Passive House Standard in 2015, which draws directly on the practice in Germany. It should be noted that sustainability practice, especially at the macro and meso levels, is contextually based and every city has its own prioritised sustainability issues to address. In developed countries, one main issue is to bring down building energy consumption and car dependency; in developing countries, by contrast, the lack of infrastructure tops the agenda.

3.2 Global Examples at City Level

3.2.1 *Freiburg, Germany*

In the field of eco-development research, Freiburg is acknowledged to be one of the very earliest initiatives of green- or eco-city projects. Freiburg is a small German city, with a population of approximately 220,000 and covering an area of 155 km^2, bordering France and Switzerland. It is in the very south-western corner of Germany and has a recorded history of at least 900 years. The city has a very strong green economy with substantial environmental progress and preservation (Gregory 2011).

As the ecological capital of Germany, Freiburg's approach towards the improvement of liveability and sustainability offer a variety of eco- or green initiatives. Some of these initiatives or projects include the improvement of eco-farming, waste utilisation facilities, hydro energy and water recycling programmes, energy-efficient buildings and green living environments, solar and green industry, and etc. The city's vision and process towards sustainable development started back in the 1970s, with major initiatives on the development of alternative-green movement and towards the later 'Local Agenda 21' that later set the city's sustainability targets and adaptation of them in policy and practice (Freiburg website). Freiburg is also well known for its progress and process of natural preservation which has established a major sustainability goal for the city's clean production and preservation. One of the main key eco-features of Freiburg is its compact planning approach that has ultimately provided pathways for 'urban agriculture, forests and community gardens, as well as excellent public transport systems and high levels of walking and cycling' (Kenworthy 2006). Moreover, the city has introduced and provided several environmental technologies, such as energy-efficient buildings, bio energy, hydro energy, renewable energy technologies and the localised management of water (Kenworthy 2006). These eco-developmental activities have therefore established Freiburg as one of the greenest cities in the world.

Freiburg's most renowned project, the Vauban Neighbourhood/ District, is one of the world's most well-known eco-development projects. Vauban's construction began in 1998 as the 'the largest car-free development in Europe' (Melia 2006) and continued offering green living style and green technologies for the new neighbourhood. Its green transit-oriented development (TOD) is described as a successful linkage between a typical TOD and green urbanism which has gained substantial attention as a sustainable development model of the future (Cervero and Sullivan 2011). Scheurer and Newman (2009) also argue in favour of Vauban, noting how it has managed to bridge the gaps between green and brown agendas. This concept is regarded as the neighbourhood's *green-brown integrated vision* (ibid.), which focuses on several key aspects of sustainability, including: buildings (to be compact, energy efficient and self-governed); waste (for the reductions of material use, embodied energy and daily consumption); open spaces (for the enhancement and integration of recreation, biodiversity of the environments and water management which is localised); transport (with priorities given to green spaces and reduction of car use/dependency); energy (with focus on renewable technologies and with two distinct objectives of centralised and distributed systems); and governance (as a bottom-up approach and with promotion of community engagement in the whole process of development).

3.2.2 *Curitiba, Brazil*

The city of Curitiba, Brazil is one of the world's most well-known early eco-city initiatives/projects. It is renowned for its advanced integrated and innovative public transportation system, which have significantly improved the environmental quality of the urban environments as well as the quality of life in the city (Macedo 2004). One of the main sustainable features is the city's extensive bus rapid transit (BRT) system, which indicates one of the key success stories of any eco-development project (Lindau et al. 2010). The future plans are for a further expansion of the BRT system and for its improvement through the use of 'advanced traffic management and user information systems' (ibid.).

The Curitiba green or eco-development project can be traced back to the early 1980s and it developed some of the early and most influential concepts of the eco-city. Located in the southern part of Brazil, Curitiba is home to almost two million people at the municipal level (and over 3.2 million people in the metropolitan region), covering an overall area of 430.9 km^2. Curitiba is the capital of Paraná and has a strong agricultural background as one of the country's key administrative and political centres. Curitiba is a major trade hub which enjoys significant environmental features such as mixed forests, botanical gardens and greenhouses as well as major water catchments, including rivers and streams. The city is also an important trade and services centre with a strong focus on developing a green economy.

Early studies of Curitiba highlight the city's relatively high average of green spaces per inhabitant as one of its major environmental advantages (Herbst 1992; Rabinovitch 1992) and the city's overall development plan has proven to indicate alternative routes towards sustainable urban development (Moore 2007). The city's planning process demonstrates a successful integrated planning approach to sustainable development, in which Curitiba's public transportation system plays a major role. The city's public transport system was developed significantly during the 1970s, when the new urban design structure was emphasised on 'linear growth along structural axes' (Rabinovitch 1992), which then enabled shaping the overall structure of the city. The low-carbon transport and green growth of the city supports the reinvention of mixed-use characteristics of Curitiba that are supported by linear corridors alongside the BRT systems in an integrated highly efficient development approach (Bongardt et al. 2010). This is effectively visible in the city's progressive approach to the integration of land use and transportation planning. Moreover, this has enabled the protection of environments, forming a bicycle path network and supporting flood control (Rabinovitch and Leitman 2004).

Overall, the integrated urban planning of Curitiba can be viewed as a successful representative model of a low-carbon eco-city. The city's environmental protection programmes have two objectives of preservation and addition, which is indicative of the strong green agenda of the city.

3.3 Global Examples at the Neighbourhood Level

3.3.1 Beddington Zero Energy Development, London, UK

Beddington Zero Energy Development (BedZED) is the largest eco-village in the United Kingdom, having been designed by architect Bill Dunster and developed by the Peabody Trust. BedZED comprises 82 homes, office space and live-work units. The village has a mix of social housing, shared ownership, key-worker homes and private houses for sale at prices that are comparable to those of more conventional homes in the area.

The key features of BedZED's planning and design are summarized as follows (BRE 2002):

Mixed-use development: A mixed-use development offers the opportunity to work locally and provides an increased sense of community resulting from the layering and interaction of different activities and occupation patterns. Building mixed-use developments at high densities can also reduce the need to develop greenfield sites.

Home Zone regulation: At BedZED, 'Home Zone' principles are designed to involve measures including reducing car speeds, giving priority to cyclists and pedestrians, having safe and convenient cycle routes and providing secure cycle storage facilities.

Public transport and car sharing: The BedZED development is located on a major road used by two bus routes, which connect to local centres. BedZED has also established a car club to reduce car ownership.

Site ecological conservation: A Biodiversity Plan was developed to maximise spaces for wildlife in the urban environment. Existing features of the site have been retained or enhanced to increase biodiversity and natural amenity value.

Energy efficiency design of the buildings: First, energy efficiency can be increased through considered zoning of activities. Second, insulation levels are considerably higher than those required by the Building Regulations. Third, triple-glazed, krypton-filled windows with low-emissivity glass, large panes and timber frames further reduce heat loss.

Energy efficiency appliances: 'A'-rated domestic appliances (including light bulbs) were specified throughout the development. Such appliances can cost a little more to buy, but return considerable energy savings and reduced running costs.

Renewable energy: The main source of energy in BedZED is a Combined Heat and Power (CHP) plant which runs on chipped tree surgery waste. The CHP has been sized so that over the course of a year it generates enough electricity to provide for all of the development's needs, which makes BedZED a zero fossil energy development.

Water efficiency: By using water efficient fittings and appliances such as washing machines, spray taps, showers and dual-flush low-flush toilets, reductions in mains water usage of 40% are achieved.

Waste disposal: Target for waste during construction was to be set at 5% of total. construction material. Recycling and composting facilities were incorporated at design phase.

Sourcing construction materials locally: 52% of the construction material (by weight) was sourced from within a 35 mile radius of the site, reducing pollution and energy impacts from transportation and to encourage local industry.

Avoiding unhealthy materials: Certain materials have been avoided due to their potential health risks to builders, occupants and future generations.

BedZED has received considerable attention since its completion and it continues to be an often-cited case study (Keeler and Burke 2009, p. 224) in discussions of sustainable neighbourhood development. The monitoring of BedZED's performance indicates that, compared to the current UK benchmarks, it secures a reduction in hot water heating of 45%, electricity consumption is 55% less, and water consumption about 60% less (Twinn 2003). In an ideal situation, a four-person BedZED household can reduce the overall eco-footprint from 6.19 hectares (typical UK lifestyle) to 1.90 hectares (Twinn 2003).

3.3.2 Hammarby Sjöstad, Stockholm, Sweden

Hammarby Sjöstad has been developed on a brownfield site, an industrial area close to Stockholm harbour and covering a total area of 200 hectares (494 acres). This sustainable neighbourhood uses a closed-loop holistic approach defined as an 'Eco-city model' at a district scale to create self-sufficient neighbourhoods. The Hammarby Model has demonstrated a "closed-loop urban metabolism, accounts for the integrated infrastructure systems of energy, water and waste from the very beginning" (Ignatieva and Berg 2014). It targets to accommodate 11,500 residential units with 26,000 population by 2017 and 20% of the total housing stock is devoted to social housing (Tsenkova and Hass 2013). This project has won international recognition, being copied around the world—e.g., in the Caofeidian Ecocity development in China (Gaffney et al. 2007).

Based on Ignatieva and Berg (2014), Gaffney et al. (2007), and Tsenkova and Hass (2013), the key features of Hammarby's planning and design are summarized as follows:

Compact with sizable green space: Hammarby Sjöstad has been planned with a dense settlement structure with typically 4–5-storey buildings in a compact neighbourhood outline, but with reasonably spacious green courtyards.

Mixed development: Hammarby provides 200,000 square metres of commercial space providing jobs for 10,000 people. The ground floors of nearly all the buildings have been designed as flexible spaces, suitable for retail, leisure or community use.

Community centre and environmental awareness: Hammarby has launched extensive efforts into educating and encouraging its residents to make full use of the project's environmental programme. The community centre—the Glass House—functions as a space to showcase technical solutions, and to advise locals on environmental issues. Residents have a display in the kitchen where they can see, in real time, how much they have used for heating, electricity and water.

100% renewable power supply: The Hammarby model includes energy conservation measures in which the goal is to reduce heat consumption by 50% and to use electricity more efficiently compared to the Swedish average. Furthermore, energy supply will be based solely on renewable sources. The electricity content will be based on solar cells, hydropower and biofuel technology. Solar panels have also been located on roof tops and solar cells cover building façades harnessing the radiation energy of the Sun and transforming it into electrical energy.

All heating obtained from heat recovery from waste combustion and sewage water purification process: All wastes from the area will be sent to the incinerator to produce both locally generated heat and co-generated electricity. Sewage water is cleaned and purified at a large sewage plant just outside the area and the waste is then recycled into natural gas. Heat produced through this purification process is then recycled for use in a district-heating unit.

Sustainable drainage design: The rainwater from surrounding houses and gardens is led by an open drain system that drains out to the attraction channel. The water then runs into a series of basins, where it is purified and filtered through sand filters or in the artificially established wetlands of the area. Roof gardens are also installed widely in the community to reduce roof run-off during storm events.

Diverse transport means: More than 95% of the residents travel to work by sustainable public transport, such as ferry, bus, on foot or by bicycle. Other features of the sustainable local transport system include networks of pedestrian and bicycle routes, and a large carpooling system.

3.4 GLOBAL EXAMPLES AT THE BUILDING LEVEL

3.4.1 *The Crystal, London, the UK*

One of London's new landmarks is home to an exhibition on sustainable development and global knowledge hub, owned and operated by Siemens. The Crystal building is a single-standing mixed-use (public and office) building, located on Royal Victoria Dock in East London. The building is part of the 'Green Enterprise District' policy in East London and is open to public as a showcase platform for the latest technologies on infrastructure and sustainable cities. The overall size of the site is 18,000 m^2. The building uses solar power and ground source heat pumps as the main means of generating energy. The Crystal is already a major iconic building of East London with several sustainable technologies that set a high benchmark for sustainability. The building was opened to the public in 2012 and is known to be the world's first building to achieve two of the highest sustainable building awards from two leading accreditation bodies, LEED and BREEAM. The crystal has achieved 'Platinum' and 'Outstanding' awards from the two accreditation bodies, respectively. The building has also achieved maximum sustainability and efficiency in terms of both building design and construction.

The crystal was designed in two crystal-shaped sections and was designed by Perkins+Will (Fit-Out, design leader) and Wilkinson Eyre Architects (shell and core design). The building and civil engineers on the project were Arup Group Limit and the exterior spaces and public realm of the building was developed by Townshend Landscape Architects. The combination of project teams indicates a project of multiple experts for one of the most sustainable public buildings in the globe. It exhibits state-of-the-art technologies for building efficiency, sustainable cities and Siemens' Environmental Portfolio.

The building site includes a public realm and the building itself. The external shape of the building leads to the creation of fascinating internal spaces, which include office spaces, conference facilities, meeting rooms and an auditorium. There are also several exhibition spaces in the building, which is an all-electric building with its own independent energy generation source. Solar power generation is strongly adopted by the building design and is major source of the building's energy production. Moreover, the building also incorporates many other sustainable technologies, such as 'rainwater harvesting system, black water treatment, solar heating and automated building management systems' (The Crystal official website). The project also incorporates several intelligent and integrated active and passive design elements, including heating, air-conditioning and ventilation systems, a weather station, lighting controls, a solar thermal hot water system, a fire alarm system, an evacuation system and a photovoltaic system.

As a result of the building's efficiency and operation of its ground source heat pumps, the building is self-sufficient for heating and cooling energy production. This energy system works with 199 pipes, totalling 17 kilometres in length, which are put into the ground at the depth of up to 150 metres. These are supported by two ground source heat pumps that generate both cold and hot water, which are then pumped back to the under-floor pipes for energy use. Cold water is then transferred to a ceiling-mounted beam which is then used for cooling purposes when needed. Moreover, the overall building energy is improved by the utilisation of 'thermal wheels'. In this process, approximately 60% of outgoing heat or cooling energy are recovered (Siemens document 2013). Furthermore, lighting and ventilation are both integrated in design and operation of the building. The building has self-shading façades which make use of high-performance solar glass technologies. This allows an overall visible light penetration of 70%, while keeping 30% of the solar energy. This particular triple-glazing solar glass includes an Argon cavity for insulation purposes. The use of such insulation systems increases the efficiency of the building energy performance. The building design also make optimal use of natural daylight in most of its internal spaces. The minimal artificial lighting system of the building uses a combination of 65% fluorescent lights and 35% LED lights with some of Siemens' advanced control and adjustment systems (e.g., detectors for dimming or switching off lights when they are not required).

The Crystal also includes a Building Energy Management System, which is one of the building's smart technologies used to detect indoor and outdoor climatic conditions. This system is utilised for control and monitoring use, energy-efficient ventilation and an intelligent lighting system. As a result, the CO_2 emissions for the Siemens offices in the Crystal are 70% lower than average for a UK office building of a similar size. In addition, water systems are among the key sustainable features of the building. Rainwater collection comes directly from the building's roof or other accessible area. The collected water is then stored in an underground storage tank and is treated by ultraviolet disinfection and filtration processes. The collected water is then recycled and used for irrigation, general cleaning and toilets across the site. It is also noted that approximately 80% of the building's hot water is heated by solar thermal water heating generated from both the roof and ground source heat pumps (Siemens document 2013).

The Crystal is a renowned building for its dual green certification. Yet the integrated technologies of the building play a major role in achieving such a status. The building, which is built on a brownfield site, utilises its surrounding conditions and adopts several energy systems and technologies. The building construction process was fully monitored to include the following three key aspects: a reduction in the use of materials; a reduction in waste production; and material recycling and re-use. In addition to the use of permeable materials for exterior surfaces, the building's green roof system also provides a key stormwater management component, which also serves as a habitat for a variety of plants and animal life. Overall, as a mixed-use building , incorporating both office spaces and public exhibition spaces, the building offers a robust platform for green building technologies and sustainable energy systems.

3.4.2 *Passive House in Germany*

Passive house is a building standard that was first developed by the Passive House Institute in Germany in the 1990s. Unlike many other green building certifications, passive house is focused on building energy performance only. It is considered to be the most rigorous energy-based standard in the design and construction industry today. Passive house has gained recognition in particular in Germany and other European countries such as Austria and France. In recent years it has expanded beyond Europe, for example, China and USA both published their own passive house standard in 2015. The main features of passive house are:

- Solar passive design, e.g., orientation, shape, window size and shading.
- Super-insulation: U-value of envelope less than 0.15 W/m^2.K to minimise heat exchange between the indoor and outdoor environments. Designing out thermal bridging is key to achieving such a low u-value.
- Super-airtightness: air infiltration less than 0.6 air change per hour.
- Mechanical ventilation with heat recovery to recover heat from indoor exhausts.
- Renewables: solar PV or ground source heat pump may be installed if the above measures cannot achieve the annual target of heating and cooling load.

Passive houses make efficient use of the Sun, internal heat sources and heat recovery, rendering conventional heating systems unnecessary throughout even the coldest of winters. Similarly, during the warmer months, passive houses make use of passive cooling techniques such as strategic shading to keep comfortably cool. In terms of performance measuring, passive houses allow for space-heating and cooling-related energy savings of up to 90% compared with typical building stock and over 75% compared to average new builds (PHI 2017). Passive house requires maximum of 10 W/m^2 for heating or cooling demand (equivalently 15 kWh/m^2.year) in residential buildings. This is a significant improvement compared to conventional residential buildings in Germany, which often employ a centralised hot water heating systems consisting of radiators, pipes and central oil or gas boilers and has an average heating load of approximately 100 W/m^2. In a passive house, traditional heating (or cooling) systems are not needed. Figure 3.1 shows an abstract model of a passive house that is well insulated and airtight. Heat exchange occurs between fresh air and indoor exhaust air in a heat exchanger, thereby greatly reducing ventilation heat gains/losses.

The world's first passive house, Darmstadt-Kranichstein, was built in the city of Darmstadt in Germany in 1991, and comprises a row of four houses, with each accommodation unit having a floor area of 156 m^2. As the first building of its kind, it was not very economical at the time because the building components had to be manufactured individually and were therefore expensive. The houses were also equipped with highly precise data monitoring devices in order to check the achievement of the objectives (Passipedia 2017).

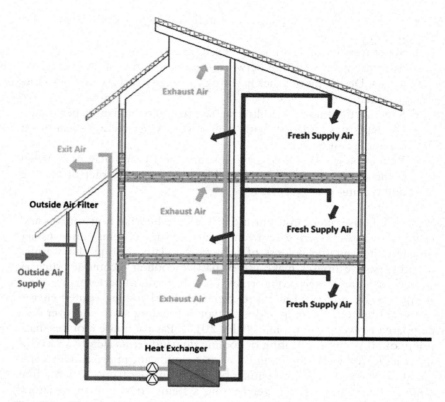

Fig. 3.1 The working of a passive house (redrawn by the authors based on the original building plan)

The house has a well-insulated envelope with u-value less than 0.15 W/m².K and an infiltration rate around 0.3 air changes per hour. The windows have three panes with low emissivity to reduce heat transfer through glass. It is installed with a mechanical ventilation system with heat recovery. Fresh air initially exchanges heat with the ground and then exchanges heat again with the indoor exhaust air before entering the indoor space. Thus the heat loss caused by introducing fresh air in cold winter is minimized.

The passive house in Darmstadt has been inhabited by four families since it was completed. The continuous performance monitoring in 2010, twenty years after its completion, indicates the measured space heating demand remains at 10 kWh/(m²a). No large maintenance measures have yet been undertaken and all building services remain unchanged from their original configuration. The façade, roof and windows remain unchanged (Passipedia 2017).

3.5 Evaluating the Sustainability Performance of the Built Environment

There is a large number of tools and approaches which have been used to evaluate environmental performance of the built environment, ranging spatially from building materials and products, energy- rated appliances, indoor air quality to whole building assessment, neighbourhoods, districts and cities. This section will give a brief review of these evaluation systems at three spatial levels: building, neighbourhood and city; that is, respectively, at the micro, meso and macro levels of the urban built environment.

3.5.1 *Green Building Evaluation*

3.5.1.1 *A Brief of Green Building Rating Systems*

Ever since the 1990s, green building rating systems have been developed to certify building performance using an assessment matrix. Cole (2010, p. 273) points out that green building rating systems (which are voluntary and market-based) have been the most important mechanism in the improvement of building performance over the past few decades. The primary objective of these systems is to stimulate market demand for buildings with improved environmental performance.

The building rating systems have directly influenced the performance of buildings. Examples can be seen in many countries around the world, including Leadership in Energy and Environmental Design (LEED) in the US, the Building Research Establishment Environmental Assessment Method (BREEAM) in the UK, the Comprehensive Assessment System for Built Environment Efficiency (CASBEE) in Japan, and Green Star in Australia. There are also signs that these assessment systems have moved beyond voluntary marketplace mechanisms by being endorsed by public agencies and other organizations as compulsory performance requirements; for example, LEED has been required by many federal agencies, and state, county and local governments as the compulsory requirement for new buildings funded by them in the US (Retzlaff 2008). In Australia, government-funded public buildings also need to meet the requirement of Green Star.

A typical building rating system is composed of a checklist of items organised into categories such as water, energy, siting, planting and indoor environmental quality, some of which may be optional. In most systems, each item is assigned a point value, and users must obtain a certain number of points in each category. The judgement of which item should be included in a system and the assignment of point values are subjective. Ultimately, a building receives a total score to reflect its environmental performance. Often, the scores are used to assign a ranking, such as platinum, gold or silver. Users typically pay to use the system, and in return they receive a variety of benefits, mainly involving market recognition and promotional opportunities. Many of the major building rating systems offer a suite of products, each of which is targeted at a specific building type, phase or situation, for example, commercial or residential, multi-unit residential or single standing residential.

The use of this type of building assessment system is not only for labelling purposes and marketing promotion. In many cases, they have also been used as planning and design tools (Cole 2005; Retzlaff 2008, 2009) because they present a set of organised environmental criteria. By default, they are understood as being the most important environmental considerations by the planning and design teams (Cole 2005). In practice, architects often set a number of certification requirements as design targets and involve green building certifiers during the course of design.

3.5.1.2 *A Further Discussion*

The sustainability of a building has internal dimensions and is also affected by factors in its surrounding environment. Retzlaff (2008) uses a scaling system of five hierarchies to examine different systems and concludes that most building-level rating systems assess performance on a fairly small scale. These systems tend to assess the building in isolation from the rest of the world (that is, in the neighbourhood and urban contexts). They are limited to building sites and are focused on building environmental performance. To some extent, this may be inadequately explained by the fact that they are whole-building assessment tools, and essentially deal with issues lying within their sites. It may also be because many of the issues related to building impacts (especially social and economic) are difficult to quantify when considered on the basis of a single building and would be more suitably addressed at a neighbourhood or urban level (Lowe and Ponce 2008).

Building rating systems vary significantly in how they were developed and how they are applied to buildings. It is interesting to note that vastly different results would be produced when different rating systems are applied to the same building. Slavid (2009) discusses an empirical case study conducted by a Glasgow-based simulation company IES. The purpose of this study was to conduct a comparison between LEED, CASBEE and Green Star. A hypothetical eight-storey commercial building in Dubai is used as the case study. The case study building failed its LEED assessment. Under BREEAM, the building fell into category B for its energy rating, which gave it two out of a maximum of 15 points available. In contrast, under Green Star, the building scored 11 points out of a potential 20. As explained by the research leader, Green Star was designed for a hot climate, and LEED covers all the very diverse US climate zones. Kawazu et al. (2005) also conducted a similar comparative study. Four high-performance office buildings in Japan and one fictitious low-performance building were evaluated using LEED, BREEAM, Green Building Tool (GBTool) and CASBEE. The assessment results showed that BREEAM and CASBEE scored higher than LEED and GBTool. Where the Building Research Establishment (BRE) in the UK evaluated the systems under normalized conditions across all the rating criteria it was found that LEED, Green Star and CASBEE assessments are not equivalent to BREEAM. Accordingly, a six-star Green Star building (the highest Green Star rating possible) is less 'green' than a Platinum LEED building (the highest LEED rating possible) and approximately equal to a 'very good' BREEAM rated building (the second highest BREEAM rating possible). BRE concluded that the four assessment systems cannot be directly compared. Generally building rating tools are all designed for internal comparison between buildings scored under the system, rather than comparisons of buildings appraised under different systems.

Variation may also occur in the same country across different climate zones. Some criteria are relatively easy to achieve in one location, but not in another. Cidell and Beata (2009) have demonstrated that spatial variation in the implementation of the LEED assessment does exist across the US. Variability across criteria and across space underscores the intuitive fact that designers, architects and builders take advantage of the flexibility allowed in the LEED certification process, and that they apply the criteria that best fit the budget, resource constraints and human and physical environments of specific projects (Cidell and Beata 2009).

In recent years, there has been a substantial increase in both the uptake of green buildings and the number of green building rating tools. For example, the World Green Building Council (WGBC) currently has 71 member councils that have given some level of commitment to green building development in their respective countries. Another 31 countries may join WGBC in the near future, as indicated in its official website (WGBC website 2017). LEED is the most commonly used whole-building rating system around the world. Between 2000 (the year of its inception) and the end of 2006, a cumulative total of 5000 projects had been registered with LEED certification across 24 countries. This number has been increased to nearly 80,000 projects across 162 countries at the end of 2015.

3.5.2 Neighbourhood Sustainability Rating Systems

3.5.2.1 Scaling up from Individual Buildings

Building-level rating systems are site-limited and mostly concentrate on environmental issues and related technologies being employed. They are incapable of addressing neighbourhood built environment, as a combination of buildings and open spaces such as roads and parking lot. Buildings are built within neighbourhoods and the spatial arrangements of neighbourhoods have significant impacts on the environmental performance of buildings and incur direct, as well as indirect, costs to households.

Green buildings are closely associated with sustainable neighbourhoods, but individually they do not make sustainable neighbourhoods. The efforts to reduce the environmental impact of individual buildings will be more or less successful depending on the opportunities and constraints of their neighbourhood's development form. For example, a mixed-use neighbourhood development can greatly reduce the use of private cars. Thus, there are incentives to expand building-level rating systems to the neighbourhood scale. Many leading international building-level tools now have a companion neighbourhood-level tool, such as CASBEE-Urban Development from Japan in 2007, LEED-Neighbourhood Development from US in 2009, BREEAM-Communities from the UK in 2012, Green Star-Precincts from Australia in 2012 and DGNB (Deutsche Gesellschaft für Nachhaltiges Bauen)-New Urban District from Germany in 2013. They have maintained a similar assessment structure and procedure, but the assessment scope is broader with focuses on neighbourhood pattern and the interrelationships between people and space.

3.5.2.2 Overview of Neighbourhood Rating Systems

Aforementioned, a neighbourhood physically encompasses multiple buildings and their sites, and the public environment such as roads, open spaces and landscaping features which exist in-between those sites. Neighbourhoods also embrace socio-economic features such as social interactions that are generally more intensive at this spatial scale. It is obvious from the foregoing discussion that, without modification, the current single-building assessments are not capable of dealing with the neighbourhood BE of merged sites and complex socio-economic environment.

Though different names are adopted for these neighbourhood systems, respectively 'neighbourhood development', 'urban development' and 'communities', they all represent a larger scale of the BE beyond single building sites. There are no confines on their application to different spatial scales. For example, LEED-ND defines a wide range of development that may constitute whole neighbourhoods, fractions of neighbourhoods or multiple neighbourhoods. There is no minimum or maximum size for single projects and no strict definition of what would comprise a neighbourhood. The current pilot projects under the LEED-ND programme range from 0.17 acres to over 12,000 acres (USGBC website). The BREEAM-Communities defines the size of developments that fall into its jurisdiction as small (up to 10 units), medium (between 10 and 500 units), and large. These scale definitions actually are open-ended in terms of spatial levels of the BE. They can be applied to any 'organized urban area' that is beyond a single building site and are universally subject to laws, planning regulations and urban masterplans.

3.5.2.3 Neighbourhood Sustainability Labelling

It is evident that both building- and neighbourhood-level assessments focus on different spatial scales, i.e., the individual building assessment concentrates on issues such as indoor environmental quality, thermal insulation and ventilation while the neighbourhood dimension of the tool emphasizes issues such as location, transport and community. An examination of LEED-ND, BREEAM-Communities and CASBEE-UD indicates that they have broader assessment scale than their building versions. On average, 55% of the total criteria examine community and urban issues and 32% address development level. Only 9% of the total criteria are limited to building site level. In general, compared to their building versions, neighbourhood rating systems are more focused on urban issues, particularly on three aspects: open space, social interaction and spatial features (e.g., location, accessibility, urban infrastructure, and so on). The common issues that are covered by LEED-ND, CASBEE-UD, and BREEAM-Communities, includes: Conservation of site ecology; Reviewable energy; Reducing water use; Reuse of rainwater and greywater; Access to local amenities and facilities; Universal accessibility; Urban integrity; Community involvement; Stormwater management; Access to public transport; Public transport capacity; Use of bicycle and electronic vehicles; Transport planning and management; Waste disposal facilities; and Construction code and green building certification. It can be seen that spatial patterns and the interrelationships between an 'area' and its urban matrix are emphasized in neighbourhood-level tools (Deng and Prasad 2010).

Similar to building rating systems, neighbourhood rating tools are regionally specific, and, thus, their focus varies across different systems. For example, LEED-ND gives more emphasis to mixed-use development such as compactness, accessibility and walkability with relatively less consideration given to local and global environmental impact, urban infrastructure capacity and the economic dimension. Density is now considered a critical component in sustainable urban development in North America and is central to many of the credits within LEED-ND (Cole 2010, p. 280). The weight of this issue does not seem to be extremely significant when compared with what is obtainable in CASBEE-UD. This reflects the fact that many Japanese cities are already built at high urban density. In such high-rise and high-density compact cities, the questions could be more concerned with air and noise pollution, limited access to daylight and natural ventilation, and limited space for vegetation. Thus, CASBEE-UD is more concerned with environmental issues such as site ecology, local environmental impact, urban infrastructure performance, and construction management. It places relatively less attention on issues such as project location, street layout, and housing affordability which are emphasized in LEED-ND.

3.5.3 *Sustainable City Rating Systems*

Since the late 1980s, there are a large variety of urban sustainability toolkits, urban assessment systems, and urban sustainability evaluation systems which have been introduced and implemented in the field of urban sustainability. Some of the most well-known ones are developed and regularly discussed by influential organisations such as City Index, the United Nations Human Settlements Programme (UNCHS), the World Bank, the Organisation for Economic Co-operation and Development (OECD), the European City Index and so on. Some of these toolkits, such as those provided by the International Union for Conservation of Nature (IUCN), the United Nations Conference on Environment and Development (UNCED) and EUROSTAT, are more focused on certain aspects like environmental sustainability (Moldan et al. 2012). These organisations have developed and for long have used their urban sustainability systems for policy development or in practice. In a holistic approach, the urban sustainability assessment systems often encompass three main pillars of sustainability (including environmental, social and economic), while some also include 'governance' as the fourth pillar. Also in most cases such assessment toolkits or systems focus on key individual dimensions, such as environmental (i.e., IUCN as stated above). Any of these approaches leads to development of framework and suggested indicators for the measurement of sustainable development. This is then applied further for policy development and practice implementation.

The measurement of sustainable development is often considered to be a shift from 'measuring economic phenomena'. At an urban scale, this is considered to be a more effective measure as macro-economic indicators (such as GDP) are often significant parts of the city performance evaluation (Joint report by UNECE/Eurostat/OECD Task Force 2013). This also leads to a lack of social development and environmental progress, which can be partly in conflict with mere economic growth. As a result, sustainable development indicators (SDIs) provide a more holistic approach towards achieving urban sustainability and pay more attention to social issues, environmental concerns, quality of life and human wellbeing. Key factors of climate change, natural preservation, resource use and environmental protection often shape the overall framework of a sustainable development assessment. This is described as an approach to harmonisation (ibid.), leading to the development of performance indicators or sustainability assessment systems that are enabled to promote progress towards a sustainable development or society.

3.5.3.1 Definition and Description
As defined by Phillips (2014, p. 6869), urban sustainability indicators are 'ways to measure the conditions and status of an urban area with a variety of factors'. The main difference between urban sustainability indicators

and other types of indicators is in their nature of integration and linkages within sustainability dimensions. Urban sustainability indicators can—to some extent—be interpreted as a method of SWOT analysis at macro or city levels, where we can identify the strengths, weaknesses, opportunities and threats that can be addressed or improved through the process of development. Hence, the system-like nature of urban sustainability evaluation methods enhances the possibilities of making corrections or suggestions for future development. This approach works in an information-based system that generates the progress and happenings of development (Phillips 2003). Rametsteiner et al. (2011) also add that urban sustainability systems are aimed at promoting 'an understanding and insight about how human and/or environmental systems operate; they suggest the nature and intensity of linkages among different components of the studies systems, and they offer a better understanding of how human actions affect different dimensions of sustainability (economy, environment, social issues)'. It is, therefore, important to keep the interconnectedness of indicators or components of key dimensions of sustainability, in order to achieve an assessment system.

An example of such approach is developed by Sustainable Measures (2010), called 'Sustainability Competency & Opportunity Rating & Evaluation (SCORE)', which demonstrates sustainability pathways for enhancement and improvement of policies and practices. Similar to other urban sustainability systems, their approach includes key dimensions of sustainability (social, environmental and economic) with sub-categories of: education, health, poverty and crime (all for social), water quality, air quality and natural resources (all for environmental), and stockholder profits, materials for production and jobs (all for economic). Although this demonstrates a relatively simplified urban sustainability system, it elaborates on the importance of linkages between the subcategories between and across the three pillars of sustainability.

Moreover, the combination of assessment indicators would then feed into a measuring system to monitor information about past trends, current realities and future direction in order to aid decision making (Phillips 2014). Also, according to Cravic (2011, p. 223), a set of sustainability indicators aims to identify issues and suggest necessary actions for resolving or eliminating the problems. This is, however, very much dependent on the availability of data and data provision for analytical and critical assessments. As a result, an information-based platform is a necessity to urban sustainability assessment systems. This is also applied in development of conceptual framework for urban sustainability assessment systems that often have two sides of theoretical application and practical implementation.

3.5.3.2 Conceptual Framework for Urban Sustainability Assessment: Theoretical and Practical

The birth of urban sustainability indicators goes back to Agenda 21, through which urban sustainability assessment was considered as an approach to support sustainability decision making, policy development and sustainable development (Munier 2004). A conceptual framework puts into place a series of sustainability indicators for better management, improvement and development cases. For instance, in its conceptual framework UN Habitat's example of 'Global Urban Indicators' (2009) indicate five key dimensions:

- Shelter;
- Social Development and the Eradication of Poverty;
- Environmental Management;
- Economic Development;
- Governance.

The above dimensions include a selection of sustainability goals and indicators that are then used in a weighting system to measure the sustainability and evaluating the conditions of the urban environments. The recent Sustainable Development Goals (SDGs) by the United Nations—Statistics Division indicates a similar approach towards the future development of urban sustainability assessment systems. As expressed by the United Nations (2015), this is 'a robust follow-up and review mechanism for the implementation of the new 2030 Agenda for Sustainable Development' which is based on a holistic framework of indicators and statistical data 'to monitor progress, inform policy and ensure accountability of all stakeholders'. In this conceptual framework, 17 action points address indicative pathways for the 2030 agenda for sustainable development. A significant attention is given to issues of climate change, environmental degradation and strong instructions.

While city-level sustainability measures are often either too broad or lack a detailed vision/plan, the implementation of such conceptual frameworks is often called into question. The sustainability of city is dependent on several factors that can be in contradiction to one another, i.e., economic sustainability vs. environmental sustainability. As a result, we can argue that urban sustainability assessment systems, while working well at the conceptual level, should be indicative toolkits or indicator systems for sustainable development and growth; e.g., eco-development for our cases here.

Unlike the other two spatial levels of meso and micro, it is difficult to implement urban sustainability assessment systems in practice. Its broad context and measures, although often quantifiable, are not easily measurable at the city scale. While the framework can provide flexibility of use, measuring sustainable development can only occur with the realm of official statistics (Joint report by UNECE/Eurostat/OECD Task Force 2013). As a result of such use, a set of policy developments can occur that would then lead to practice implementation at a smaller scale. However, some particular components of climate change control, environmental protection and economic growth can be applied and supported at the city scale.

In the past two decades, we can track a steady increase of popularity in the development, utilisation and application of urban sustainability systems. This is significantly apparent in research and is, in recent years, more applied in policy and practice. We can also witness more involvement from various organisations and stakeholders in the development and utilisation of urban sustainability systems that are often indicators-based and are used for the assessment of city performance. This signifies the importance of such systems, benefitting substantially from city planning and urban development perspectives. Hence, the push towards policy development at the city level can play a major role in development of urban sustainability assessment systems.

3.6 How Do Global Developments Inform China?

Before we move on to the context of China, it is important to elaborate to what extent global practices and policies previously have, and continue to have, an impact on Chinese practices and policy reforms. Although this chapter has served as our entry point to the concept of eco-development from the global perspective, it also enables us to highlight some of the global–local linkages. It is important to recognise these linkages and evaluate the past and current directions in China. It is also important to note that China has taken a journey of three decades in less than a decade, if we compare China with the developed countries of the West (Cheshmehzangi

2017). The transitions have been immense and in many ways impactful on policy development and practices. This journey is what we can simply refer to as the initial phase of green/eco-development in China. It also appears as a steep learning curve for many stakeholders and actors, who have been involved in, or have taken the leading role in, developing green/eco-projects in China. This initial stage has enabled China to first learn from the global examples and then gather the knowledge gleaned for future practices and policy development/reforms in China. Among these early activities were project visits that were taken by governmental and city officials, practitioners and policy makers who visited and learned from successful, or if not, well-known models of green/eco- projects around the globe. These visits were the breakthrough for Chinese counterparts to observe, evaluate and learn from green practices. Also during this time, many Chinese (including returning overseas Chinese) and international scholars and practitioners have actively studied global examples and proposed for methods and approaches that were later adapted in the context of China. In this process, scholarly findings have grown up and matured significantly, enabling many platforms of research and development, between research, industry and policy making. For instance, the authors of this book are both examples of such scholars whom have continuously worked on the theme of green and eco-development in the last decade in China.

In addition, China has benefitted from many international relations, cooperative and collaborative projects, such as joint research collaborations with global and leading institutes, learning from the know-how and experience of examples and pathways that have previously gained global attention, if not recognition. At first, perhaps we can argue that many of these examples were not entirely adapted to the local context, but were mostly replicated without the comprehensive inclusion of local characteristics, local varieties and local values. Many aspects, such as rich local and historical characteristics, diverse climatic regions, and the top-down willingness to change the business-as-usual cases enabled China to delve into the policy mobility transfer and considering changes in practice. As a result, this trend of replication has weakened gradually over time, enabling many scholars, practitioners and policy makers to seek for localised models and the local refinement of successful examples. This also brought a chance to look back at some of the existing practices in China, of which the vernacular examples are of high popularity to many researchers who study the sustainability characteristics of vernacular forms, models and building design in China. What we anticipate seeing in future is the increase in bilateral and multilateral platforms and arrangements that will enable China to be in mutual communication platforms with project holders and developers of the other countries. China is also expected to take a more active role in the global arena, some of which are already perceptible in major global platforms, such as the well-known Belt and Road Initiative (BRI). The nature of some of these platforms will turn into south-to-south cooperation projects (as it has already instigated), enabling China to export some of its successes to the other regions of the developing world. We also foresee more international cooperation in this route, not only for China to learn from others but to share their experiences globally.

Although the progress has been very tangible, it is believed that China has a long way to go on its journey to become a green nation. Eco-development projects of China are mostly new; collectively, however, they do represent a remarkable achievement of the country's success in its first phase of promoting green/eco-development. The global examples have certainly informed what we can refer to as the first batch or first movement of green/eco-practices in China. The move towards policy reforms and the development of new policy frameworks is already seen in many regions, starting, of course, from the more prominent cities, particularly in the cities of first and second tiers. Some of these examples are further developed as joint venture projects that we see shaping up in many parts of the country. These examples are unique in their nature, experimental in purpose, and effective as large-scale and open laboratories for green/eco- or sustainable development. Several of these examples will be discussed further in the forthcoming chapters (see Chaps. 5, 6, and 7), where we introduce 21 case studies of Chinese projects at the three distinct spatial levels of the built environment.

To summarise, we can argue that China has clearly benefitted from the global examples—some may be direct through policy transfers and then transitions, and practice adaptation; and some are indirect through the review and evaluation of lessons learnt from the others. This benefitting will continue to inform China's own agenda for green/eco-development, but the important question is: will China possibly inform the global context in the future? In the chapters that follow, we will highlight some of the existing examples and models in China and will then focus on how the country will potentially be one of the future global and successful models itself.

REFERENCES

Bongardt, D., Breithaupt, M. and Creutzig, F. (2010) Beyond the fossil city: Towards low carbon transport and green growth. In 5th regional environmentally sustainable transport forum in Asia, United Nations Centre for Regional Development, Bangkok, Thailand.

Cervero, R. and Sullivan, C. (2011) Green TODs: marrying transit-oriented development and green urbanism, *International Journal of Sustainable Development & World Ecology*, Volume 18, Issue 3, Special Issue: Multidisciplinary Perspectives on Sustainable Development, pp. 210–218.

Cheshmehzangi, A. (2017) *Smart Eco Cities: Project Findings from China*, project report presentation at the International Workshop on 'Smarter and/or Greener?', Ningbo, China, 11–12 December 2017.

Cidell J. and Beata A. (2009) Spatial variation among green building certification categories: Does place matter? Landscape and Urban Planning 91, pp. 142–151.

Cole R.J. (2005) Building environmental assessment methods: Redefining intentions and roles, Building Research and Information, 35(5), pp. 455–467.

Cole R.J. (2010) Environmental assessment: shifting scales, In Edward Ng (eds) Designing high-density cities for social and environmental sustainability, pp. 273–282, Earthscan London.

Cravic, B. (2011) Integrating tourism into sustainable urban development: Indicators from a Croatian coastal community. In M. J. Sirgy, R. Phillips, & D. Rahtz (Eds.) Community quality-of-life indicators: Best cases V. Dordrecht: Springer.

Deng, W. and Prasad, D. (2010) Quantifying Sustainability for the Built Environment at Urban Scale: A Study of Three Sustainable Urban Assessment Systems, paper presented at 2010 Sustainable Building Conference South East Asia, 4–6th May 2010 Malaysia.

Gaffney, A., Huang, V., Maravilla, K. and Soubotin, N. (2007) Hammarby Sjostad Case Study | CP 249 Urban Design in Planning 2007, http://www.aeg7.com/assets/publications/hammarby%20sjostad.pdf, access 08/06/2016.

Gibson, R. B., Alexander, D. H. M. and Tomalty, R. (1997) Putting Cities in their place: Ecosystem-based Planning for Canadian Urban Regions, pages 25–40, in Reseland, M. (ed.) *Eco City Dimensions: Healthy Communities, Healthy Planet*, Gabriola Island and New Haven: New Society Publishers.

Global Urban Indicators – Selected statistics (November 2009) Available at www.unhabitat.org, accessed on 08-June-2016.

Gregory, R. (2011) Freiburg: Green City, as part of Eco Tipping Points project, also available at: http://www.ecotippingpoints.org/our-stories/indepth/germany-freiburg-sustainability-transportation-energy-green-economy.html.

Herbst, K. (1992) Brazil's model city, *Planning Chicago*, Vol. 58, pp. 24–24.

Ignatieva, M. E. and Berg, P. (2014) Hammarby Sjöstad – A New Generation of Sustainable Urban Eco-Districts, The Nature of Cities. Available at: http://www.thenatureofcities.com/2014/02/12/hammarby-sjostad-a-new-generation-of-sustainable-urban-eco-districts/ access 08/06/2016.

Joint Report by UNECE/Eurostat/OECD Task Force on Measuring Sustainable Development (May 2013) Framework and suggested indicators to measure sustainable development. Available from: https://www.unece.org.

Kawazu, Y., et al. (2005) Comparison of the assessment results of BREEAM, LEED, GBTOOL and CASBEE, The 2005 World Sustainable Building Conference, Tokyo.

Keeler M. and Burke B. (2009) Fundamentals of integrated design for sustainable buildings, John Wiley & Sons, Inc. New Jersey.

Kenworthy, J. R. (2006) The eco-city: Ten key transport and planning dimensions for sustainable city development, Environment and Urbanization, vol. 18, no. 1, pp. 67–85.

Lindau, L. A., Hidalgo, D. and Facchini, D. (2010) Curitiba, the Cradle of Bus Rapid Transit, Built Environment, Volume 36, Number 3, pp. 274–282(9).

Lowe C. and Ponce A. (2008) An international review of sustainable building performance indicators & benchmarks, UNEP-SBIC Report.

Macedo, J. (2004) City Profile Curitiba, *Cities*, Vol. 21, Issue 6, pp. 537–549.

Melia, S. (2006) On the Road to Sustainability: Transport and Carfree Living in Freiburg, document in Faculty of the Built Environment at UWA, Bristol, UK.

Moldan, B., Janoušková, S. and Hák, T. (2012) How to understand and measure environmental sustainability: Indicators and targets, *Ecological Indicators*, Volume 17, pp. 4–13.

Moore, S. A. (2007) Alternative routes to the sustainable city: Austin, Curitiba, and Frankfurt, Lexington Books.

Munier, N., (2004) Multi-criteria environmental assessment: a practical guide, Dordrecht: Kluwer Academic Publishers.

Phillips, R. (2003) Community Indicators, Chicago: American Planning Association.

Phillips, R. (2014) *Urban Sustainability Indicators*, In *Encyclopaedia of Quality of Life and Well-Being Research*, Netherlands: Springer, pp. 6869–6872.

Rabinovitch, J. (1992) Curitiba: towards sustainable urban development, Environment and Urbanization, October 1992 vol. 4 no. 2, pp. 62–73.

Rabinovitch, J. and Leitman, J. (2004) Urban Planning in Curitiba, in Wheeler, S. and Beatley, T. (eds.) Sustainable Urban Development Reader (Routledge Urban Reader Series), Oxon, Routledge.

Rametsteiner, E., Pulzl, H., Alkan-Olsson, J. and Frederiksen, P. (2011) Sustainability indicator development – Science or political negotiation? Ecological Indicators, Volume 11, pp. 61–70.

Retzlaff R. C. (2008) Green building assessment systems: a framework and comparison for planners, Journal of the American Planning Association, Vol. 74, No. 4, pp. 505–519.

Retzlaff R. C. (2009) The use of LEED in planning and development regulation, Journal of Planning Education and Research, Vol. 29, No. 1, pp. 67–77.
Scheurer J. and Newman P. (2009) Vauban: A European Model Bridging the Brown and Green Agendas, in UH Habitat Global Report on Human Settlements. Available at: http://www.unhabitat.org/downloads/docs/GRHS2009CaseStudyChapter06Vauban.pdf.
Slavid R. (2009) BREEAM, LEED and Green Star: Who's the fairest? http://www.bsdlive.co.uk/story.asp?storycode=3146922.
Tsenkova, S. and Hass, T. (2013) Planning Sustainable Communities: Europe's new model for Green living in Stockholm, Calgary, Canada: University of Calgary.
Twinn C. (2003) BeDZED, THE ARUP Journal 1/2003.
Vodden, K. (1997) Working Together for a Green Economy, pages 80–94, in Reseland, M. (ed.) *Eco City Dimensions: Healthy Communities, Healthy Planet*, Gabriola Island and New Haven: New Society Publishers.

3.6.1 Websites

BRE (2002) Case study report of Beddington Zero Energy Development, Building Research Establishment, http://www.bioregional.com/newsviews/publications/bedzedbestpracticereportmar02/.
Crystal website: https://www.thecrystal.org/about/architecture-and-technology/. Siemens, Document on The Crystal: one of the most sustainable buildings in the world, Available at: https://www.thecrystal.org/wp-content/uploads/2015/04/The-Crystal-Sustainability-Features.pdf.
Freiburg website: Document of Approaches to Sustainability: Freiburg. Available at: www.freiburg.de/greencity.
Passipedia (2017) The world's first Passive House, Darmstadt-Kranichstein, Germany, Available at: https://passipedia.org/examples/residential_buildings/multi-family_buildings/central_europe/the_world_s_first_passive_house_darmstadt-kranichstein_germany.
PHI (2017) Homepage of Passive House Institute; Available at: http://passivehouse.com.
Siemens Document: 'LEED Platinum certification for the Crystal' (press release in 2013); Available at: http://www.siemens.com/press/en/pressrelease/?press=/en/pressrelease/2013/infrastructure-cities/buildingtechnologies/icbt201311093.htm.
Sustainable Measures (2010) Indicators of Sustainability. Available at: http://www.sustainablemeasures.com/indicators.
United Nations (2015) Sustainable Development Goals (SDGs). Available at: https://sustainabledevelopment.un.org/sdgs.
WGBC (2017) World Green Building Council Website; Available at: http://www.worldgbc.org/.

CHAPTER 4

Eco-Development in the Chinese Context

4.1　China's Urbanisation Era

China is undergoing the largest scale of urbanisation in history and at an unprecedented pace. Between 1991 and 2012, China's urban population has increased from 26.4% to 52.6% in percentage terms. The urban built areas have expanded from 12,856 to 45,566 square kilometres over the same period, an increase of 3.5 times in about two decades (China Statistical Yearbook 2013). Enormous new buildings around the country have been constructed to accommodate the increased population. In recent years, China has been adding about 1.7 billion square metres of new floor space on an annual basis (Li and Shui 2015). As illustrated by the Building Energy Conservation Centre (BECC) of Tsinghua University, the annual rate of new construction in China equals the total amount of new buildings in all developed countries (BECC 2009). Taking the annual addition to the residential building stock, the number has soared from 0.4 billion m^2 to approximately 0.7 billion m^2, a growth rate of 69% in the last decade.

In March 2014, China revealed a new blueprint to further expand urbanisation, which has been referred to as China's new urbanisation plan or strategy. For the first time, this blueprint includes and integrates debatable matters of human and social developments that are critical to China's growing urban population (Cheshmehzangi 2014). Despite the delay in developing such an urbanisation plan, the new strategies include fairer

© The Author(s) 2018
W. Deng, A. Cheshmehzangi, *Eco-development in China*,
Palgrave Series in Asia and Pacific Studies,
https://doi.org/10.1007/978-981-10-8345-7_4

conditions for rural-to-urban migration. The focus is a human-centric approach to urbanisation with a major emphasis on not only the urban but also the rural areas. The possibilities of giving access to public services and urban welfare are new in discussions and are opened up for gradual development; i.e., allowing the rural residents to migrate to cities in a gradual pace, including better benefits to access services. Having said this, there remains strong consideration about rural-to-urban migration with the enforcement of physical infrastructure development and industrial growth. The predicted figures for China's urban and rural population indicate China's consistent, but major influx of rural migrants to the cities. This is due in part to China's lack of investment and development in the rural and remote areas, leaving substantial social, cultural and economic gaps between rural and urban residents. This 'uneven development' (Smith 2010) is effectively a polarisation in capital, investment and income that can be argued as a global matter, yet is more significant for the case of China as the pace and scale are unprecedented.

Furthermore, the lack of public services and growing demand for the provision of daily needs in China's rural areas are becoming more apparent while the physical infrastructure is improving at the national level. Similar to many other developing countries, there is a perceptible decline in the rural areas of China, where social and economic attributes are more vulnerable. Therefore, the increasing influx of population to cities, if further characterised by a lack of structure and undetected mobility patterns, will result in severe challenges to the course of sustainable development.

Of all concerns about the role of the social sciences in urbanism in the case of China, we can refer to strategic plans that are shaping the course of national, regional and local development. It is certainly now clear that overspending on the physical infrastructure does not necessarily result in sustainable human development. A comprehensive socio-economic approach to development seems essential to the current phase of China's urbanisation. As a result, China is now expected to move towards a more structured approach bringing in together issues of urbanisation and urbanism. Therefore, it is essential to create opportunities for both the rural and the urban in order to reduce the burden from issues of rural-to-urban migration and approach towards human development. This complexity needs to get resolved along with the increasing public demand and the increasing gaps between the social classes. However, 'class struggles of some sort' is inevitable in such a process (Harvey 2013, p. 115). Strategically, the move towards human development and humane scale approach to urbanisation requires huge force in bringing in disciplines in the social sciences and humanities. Although China is becoming a global role model as a result of its vibrant and thriving economic growth, it is still struggling with its own social and cultural aspects faced with rapid transformation(s) and even decline. China's ageing society and the growing demands from its rapidly increasing middle-class population remain as major concerns that may not allow China to be Asia's number one economic power (Cheshmehzangi 2014).

Given the current stage of China's growth and development, there is a certain need for urbanism with the support of the social sciences in making it sustainable. The change and reform in policies and development are the major mechanisms to address this need. For instance, the current urban-biased economic policies and the widespread extent of urban–rural inequalities, as discussed by Lu and Chen (2006), are issues that can only be resolved through the development of comprehensive social security and structured strategic plans. The disparities between China's rural and urban populations are not only limited to economic policies but also related to the urban–rural income gap (Yao 2005), the levels of investment and education (Wan et al. 2006) and labour allocation matters (Yang and Zhou 1999). Taking into consideration of the income disparities between rural and urban populations, we can refer to Lu's study (2002) on how this affects growth, allocative efficiency and local growth welfare; none of which can be resolved without a considerable input from the social sciences. In this respect, Yao (2005) discusses that a 'non-equilibrium financial development in China may lead to negative economic consequences', resulting in even larger gaps between different social classes and backgrounds in the country (mainly between the rural and the urban). Hence, some unprecedented challenges are currently escalating in China as they become worrying indicators for economists, local governments and policy makers. Some of these challenges have been under discussion and debate over recent few decades, including: changes in labour productivity between different sectors (Nolan and White 1984); issues of public policy on migration (Wong 1994); the threshold effect of income inequality on China's real level of economic development (Wang and Ouyang 2008); and impacts on urbanisation of the heavy-industry-oriented development strategy (Lin and Chen 2011). Alongside these already itemised challenges, the author aims to explore China's current and forthcoming challenges as part of the new urbanisation phase.

4.2 Towards a Comprehensive Development Model

4.2.1 China's Challenges of Urban Growth and Urbanisation: Lessons for the Future

China's new urbanisation is currently facing three major challenges: (1) rural decline under conditions of increased urbanisation; (2) over-urbanisation in coastal areas; and (3) unbalanced development for small and medium-sized cities (from the conclusion session of the EU–China Event on Urban Innovation 2013, Foshan, China). These three challenges bring about social pressures, such as social inequalities and social decline, calling for the development of social sustainability. The move towards one billion urban residents by 2030 (McKinsey 2009) indicates substantial pressure on China's future urban growth and urbanisation. Therefore, China's trend for urbanisation will require a more holistic approach than has taken place since the 1980s (Cheshmehzangi 2014).

The opening speeches by several mayors of Chinese cities in the major 'EU–China Urban Innovation Event' in May 2013 indicate three overlapping aspects: (1) People; (2) Land; and (3) Income. These three elements are the main pillars of the new phase for China's urbanisation. People, central to the current reforms, can boost China's growth by increasing the consumption. This is simultaneously a major challenge for local governments with pressures on expenditure and further demands for investment. However, in the coming years the main challenge will remain on development of mechanisms for the new urban residents, principally those who have migrated from rural areas. Through the gradual reform of the Hukou system (i.e., the official record of household registration in a government system in China), it is anticipated that a better balance should be achieved between living qualities, social development and migration. Similarly, land will play a significant role in China's new urbanisation. Cities will have to develop a better scientific understanding of urbanisation and urban growth, enabling them to become more inclusive, prosperous and international. The internationalisation of major cities and regional urban competitiveness will remain vital to development of thriving urban hubs in the country. Yet the forthcoming reforms should include substantial planning changes, such as: land use reforms (e.g., land ownership in particular); the integration of land issues with growing environmental concerns (e.g., global warming, exploitation of resources and pollution); and agricultural production and ecological matters (e.g., production and efficiency of food supply and ecological protection). Lastly, the issue of investment and how income is generated and raised in comparison with the global figures will remain a principal aspect of China's new process of urbanisation. The major demand will be for institutional innovation and the development of public services, realising a balance for development of a fair pricing mechanism. The marketisation for both local and international markets will remain imperative in indicating how some of the urbanisation issues can be resolved over time. To sum up all three aspects, the forthcoming actions will require focusing on understanding people's needs, investing in social development and achieving a more harmonious approach to urbanisation. The pace of urbanisation may fall as the above challenges become more significant at certain times and/or certain parts of the country. The approach to urbanisation will require a holistic understanding to be achieved at both macro and micro levels, introducing considerable reforms, and supporting both communities and the local governments. The lessons learnt from the past will prioritise major aspects of environmental, agricultural and ecological importance in urbanisation and will be embedded in the process of urban growth (Cheshmehzangi 2014).

4.2.2 *Urbanism and Urbanisation: Towards a Comprehensive Development?*

Back in 1993, Nicholas Kristof (1993, p. 59) argued that 'the rise of China, if it continues, may be the most important trend in the world for the next century'. China is now establishing itself as a successful model for other developing countries, particularly for the rapidly developing countries. Yet it is often forgotten that urbanisation and development should not only be replicated but also require careful consideration in terms of how it is utilised, localised and reflected on the local demands. In this sense, we can refer to the role of urbanism in the process of urbanisation as a key matter. Through both governance and increasing the efficiency of planning and development, it is important to concurrently develop the welfare system and infrastructure, allowing for innovative and integrated solutions to overcome issues of poverty, deprivation and even decline. Urbanism, if it is to be integrated in this process, can help to promote the quality of urban development patterns and help to develop better institutional reforms and measures for an improved quality of life. It can then help to achieve a more sustainable way of development and growth. The follow-up matters will gradually shift towards the improvement and/or enhancement of public services and the management of social development, both of which are essential to the achievement of a sustainable urbanisation.

As argued by Harvey (1985, p. 23), the relation between 'city formation and the production, appropriation and concentration of an economic surplus has been long noted'. Therefore, the dynamics of urbanisation are interlinked and often overlap. For China's new urbanisation process, the formation of a holistic approach is particularly critical, questioning how and to what degree it can happen in the coming years. The increase in China's production and demand are yet to bring more success and failed stories. In the current phase of new urbanisation, it is vital for new policies to enable the formation of cities beyond the current structure. The aim should ultimately be towards structuring and developing socio-economic strategies. This will potentially offer a mechanism to promote a less pro-planning urbanism in urbanisation, and may instead provide a comprehensive approach to shape a framework for future growth and development. Moreover, the successful effect of urbanism in urbanisation requires significant support and input from a wide range of disciplines in the social sciences. We can argue that a multilayered approach to urbanism is required to ensure the achievement of a comprehensive development. This cannot happen solely on the basis of economic growth or social development. The combination of the two, alongside environmental concerns and cultural values, are essential to achieving successful models. In this respect, we can simply refer to Jane Jacobs' statement (1961) that the combination of decay in cities, decline in economy and intensifying social distresses/pressures are not coincidental. The impact from one on another is becoming so apparent that it is often challenging to separate them from one another.

In the case of China, we can point out how the continuous social struggle and unbalanced economic situation are intertwined. The enduring issues related to rural migrants and the continuous decline of rural communities raise alarm bells of how this trend may continue in future. Undoubtedly, there will be more pressure on policy makers, as such major matters become more challenging. In this process, if the role of urbanism is neglected we may end up with policies and decisions that diminish the content of urbanisation. The content, embracing development and growth, will require careful consideration to become nourished in the process of urbanisation. On the other hand, there are major concerns on environmental implications of urbanisation and lifestyle changes in China (Hubacek et al. 2009), which again reflect on inequalities in income and lifestyles, especially between rural and urban populations. From a different perspective, this is also discussed by Zhang and Song (2003) through their cross-sectional analyses on rural–urban migration and the impact on urbanisation. In their study, Zhang and Song (ibid.) refer to the rural–urban income gap as a major challenge for the process of urbanisation. Although their paper pays attention to the issue of inequalities, the authors' place an emphasis on considering the rural as the pivot point of urbanisation. This merely means we cannot neglect the impact of urban growth (both physical and economical) on the rural communities. We also should not allow the significant decline of rural and peri-urban areas in the process of urbanisation. Therefore, carefully crafted planning is required to consider rural, peri-urban and urban areas under a single umbrella of comprehensive development. This will need to include contextualised 'development policies and institutions' (Whyte 2009, p. 371) and consider all previous development patterns and opportunities of future growth and development.

4.3 Evolution of Eco-Development in the Chinese Context

The term eco-development is a relatively one, although it can be tracked back to some ancient Chinese philosophies, as discussed earlier in earlier chapters. It originated first in the context of building energy saving in China. The Chinese central government began to take measures to reduce the building energy consumption in 1986 when the Domestic Building Energy Saving Standard (JGJ26-1986) was implemented and an energy-saving target of 30% (compared with the benchmark buildings built in 1980) was established. A milestone of eco-development at the building level is the launch of the Green Building Evaluation System (GBES) in 2006, which is the first national standard to evaluate comprehensive building performance in China, rather than focusing only on energy efficiency.

Eco-developments at the neighbourhood and city scales have occur much more recently. The concept of 'eco-city' or 'eco-neighbourhood', for example, was not translated into practical initiatives until the early 2000s. However, the country has seen a booming of eco-developments at neighbourhood and city levels since 2008, when the Scientific Concept of

Development was unveiled by the then President Jingtao Hu at the 17th CPC (Communist Party of China) National Congress. This concept called for a development model that places people first, and aims to achieve comprehensive, coordinated and sustainable development. After that, a number of pilot projects have been launched and evaluated. The following section reviews the national five-year plans (FYPs) from the 11th to the 13th (2006–2020) which marked a time period during which China commenced a more critical and thorough reappraisal of its developmental priorities and strategies in cities.

4.3.1 Policies on Eco-Development in the National Five-Year Plans

Since 1953, China has implemented a series of Five-Year Plans (FYP) which established the blueprint and targets for national economic and social development for a five-year period. FYP provides a main window to observe the directions, strategies and changes in development mode over a five-year period. They usually include a national-level masterplan that set key national overall targets and a number of special programmes that address sectorial targets and are usually drawn up by ministries that supervise a specific sector. For example, the Ministry of Housing and Urban-Rural Development (MoHURD) is responsible for the special programme of the construction industry, and the Ministry of Environmental Protection (MEP) focuses on the special programme for environmental protection. The national overall targets are also disaggregated and allocated to provincial level. Provincial governments need to make their own five-year plans (FYPs), and the provincial targets are then broken down further to city and county levels. Therefore, governments at various levels are obliged to achieve these targets. The national FYP and its special programmes, as well as local FYPs, constitute a systematic mechanism to move the country towards the targets established.

Traditionally, FYPs have been focused on economic development and related growth targets, with less consideration on the social and environmental aspects. In an address to the National Congress in 2006, Ma Kai, the then Minister of National Development and Reform Commission (NDRC) and one of the current Vice Premiers, admitted that the accomplishment of the 10th Five-Year Plan was at the expense of resources and environment (NDRC 2006). Ma (NDRC 2006) further remarked:

During the 11th Five-Year Plan period, we will implement the basic national policy of resources conservation and environment protection, develop cycling economy vigorously, protect and restore ecosystem and environment, strengthen environmental protection, improve resources management, promote the balanced development of population, resources and environment to realise sustainable development.

Recognising the limitation of the development mode featured with intensified resource input, the Eleventh Five-Year Plan for the period of 2006–2010 was rectified by the National Congress in 2006, which was described as 'revolutionary', 'a watershed' and 'of turning point significance' (Fan 2006) and 'historic for its action on climate change' (Ye 2011) by incorporating a number of indicators addressing environmental deterioration, resource depletion and social inequity. Two basic principles were reflected in the plan: 'concept of scientific development' and 'harmonious socialist society'; in other words, economic growth was not to be sustained at the cost of environmental degradation and resource depletion, meaning that disadvantaged groups and less developed regions share the fruits of economic growth (Ye 2011).

Compared to the 10th Five-Year Plan, it is the first time that a quantitative target for energy efficiency was put in the national FYP, which required a reduction of 20% from the 2005 benchmark. This is measured as energy consumption per unit (10,000 RMB, approx. US$ 1450) Gross Domestic Production (GDP). It was also defined as one of the 'restricted' targets which are tied in with governments at all levels from the central government down to the provincial, city, county and township levels. In other words, governments at all levels have responsibilities to achieve the disaggregated targets allocated and, as a result, to contribute to the accomplishment of the overall national target. This was quite an ambitious target because the increment of energy intensity was greater than the GDP growth rate during the period of the 10th Five-Year Plan (Lin et al. 2008). The introduction of such a target signalled a shift towards a new development mode that tries to make a balance between economic growth and ecological conservation. The 20% energy-intensity target also translates into an annual reduction of more than 1.5 billion tonnes of CO_2 by 2010 (Zhang et al. 2011) compared with the 'business-as-usual' development mode.

In September 2006, the State Council issued a disaggregating scheme to provinces with quantitative targets (The State Council 2006). The highest goals are for Shanxi and Inner Mongolia, both of which have strong coal-mining industries, with a 25% reduction compared to 2005. The lowest target, a 12% reduction, was given to Hainan and Tibet. Twenty provinces out of a total of 33 were given a 20% energy-saving target, the same as the national target. These provincial targets were further broken down to the level of cities and counties. For example, the Shanxi Government approved a scheme in November 2006 disaggregating the provincial energy-saving target (25% reduction) into subtargets for each city (Zhang et al. 2011).

In June 2011, China announced that it had essentially met the energy-saving target, with a final achievement level of 19.1% (NDRC 2011). A field survey of local governments and enterprises conducted by Tsinghua University also confirmed the national reduction in energy intensity is in line with the 20% target (Ye 2013). This referred to a reduction of 1.55 billion tons of carbon dioxide emissions (ibid.).

The building sector was seen as one of three key sectors for potential energy saving in the 11th Five-Year Plan, along with power generation and the industrial sector. During this period, China had initiated a number of laws, regulations, policies, standards and pilot schemes to reduce energy consumption in buildings, including:

- The Green Building Evaluation System (GBES) in 2006, which is a national standard and is the first comprehensive green building rating tool in China;
- The Renewable Energy Law in 2006, which encourages the integration of solar PV in building design;
- The Ordinance of Energy Saving in Civilian Buildings in 2008, which is seen as a significant move in China's move to cut energy use. This establishes a framework for building energy saving by introducing various policy tools, e.g., energy-saving planning, a financial incentive mechanism, and market-access requirements for new buildings. The regulations have specified the responsibilities of designers, real estate developers, housing quality supervisors and even leaders of public institutions for saving energy in both residential and office buildings;
- Design Standards for Energy Saving in Residential Buildings in Severe Cold and Cold Zone in 2010;
- Design Standards for Energy Saving in Residential Buildings in Hot Summer and Cold Winter (HSCW) Zone in 2010.

All of these regulations, standards and policies mark China's remarkable transition to a more environmentally-friendly built environment. By 2010, 113 building projects have been labelled by GBES (MoHURD 2012). However, the efforts have been largely focused on energy conservation at the single building level, with little consideration given to larger spatial levels such as city and community. Green building development is still dispersed across the country and has not appeared at a large scale of implementation.

The 12th Five-Year Guideline was approved by the National People's Congress on 14 March 2011. The plan reiterated 'scientific development', and emphasised 'higher-quality growth', which gave priority to tackling sustainability issues, such as pollution, energy-intensive consumption, and resource depletion. It has put forward three key themes: economic restructuring; social equity; and energy and environment. This was the first time that a carbon reduction target was set in the main plan, building upon China's pledge to reduce carbon intensity by 40–45% by 2020 based on 2005 benchmarks, which had been made at the United Nations Climate Change Summit in Copenhagen in 2009 (China Dialogue 2011). The plan required a 16% and 17% reduction of energy consumption and carbon emissions per GDP unit, respectively, from the 2010 level (with 7% GDP growth). Though the energy reduction target was less than the previous FYP, it was still believed to be 'ambitious' as the easy-hit approaches such as shutting down small power plants and steelmakers have been largely implemented (ibid.). Further reductions would be more intensified on two key factors of technology and investment.

In May 2012, a Special Program for Building Energy Conservation was published by MOHURD. It is also the first time that such a sectorial plan was put forward specifically on the concept of sustainable built environment. The program aimed to promote large-scale green building development. Furthermore, in addition to promoting building performance, the program also focused on developing eco-developments at large spatial levels. It was required to establish 100 eco-cities, eco-districts or eco-industrial parks across the country for demonstration purposes, which should address issues, including:

- Energy reduction—including energy-efficiency design standards, renewable energy generation, energy consumption quota, etc.;
- Land use—including compact land use, neighbourhood pattern, use of underground space, etc.;

- Transport—including public transport network, clean fuels, green trips, etc.;
- Green building development—including a green building ratio, savings for energy, land, water, materials in buildings, indoor air quality, green construction, etc.;
- Environmental quality—including air pollution, water pollution, noise, urban island, etc.;
- Social harmonisation—including people's living quality, accessibility to facilities, community involvement, etc.

According to the official announcement by the National Development and Reform Commission, China has exceeded the targets set for the 12th FYP period. The energy intensity and carbon intensity have been decreased by 18.2% and 20% from 2010 level (NDRC 2016).

In 2015, China's urban population exceeded its rural population for the first time in its history. In recent years, urban problems such as air pollution and traffic congestion have caused widespread discontent among Chinese people. In December 2015, the CPC and the central government convened its first Central Urban Work Convention in 37 years since the reform and opening-up policy was initiated in 1978. The Convention calls for the following goals to be met in Chinese urban development (China Daily 2015):

- To enable people in central and western areas to benefit from the development of cities without leaving their hometown;
- To keep the distinctive landscape and cultural and architectural identities of cities;
- To carry on the historical and cultural heritage of cities;
- To build cities with beautiful natural landscapes;
- To keep the expansion of cities within planned boundaries and to build smart and compact cities.

On the heels of the convention, the Central Committee of the CPC and the State Council issued Several Opinions on Enhancing Urban Planning, Construction and Management in February 2016. These policy suggestions reflected the revised thinking of the new leadership, under President Xi Jinping, with regard to urban development models. Among the proposed measures are some unprecedented proposals that have stirred up widespread debate in the country. Typical urban development models over the past few decades featured with superblocks, gated small residential districts, and giant streets have been overturned. Urban planning and development are now expected to follow principles such as the following (China Daily 2015):

- Increase street network intensity and walkability;
- Open up gated communities;
- Expand mixed-use development rather than the current functionality-based zoning approaches;
- Public transport-oriented urban development;
- Improve low-carbon city technologies and "sponge" city technologies.

One month later, in March 2016, the 13th FYP was approved by the National People's Congress. The reduction targets for energy intensity and carbon intensity have been set for 15% and 18%, respectively. For the first time PM 2.5, a type of dangerous fine particulate in the air mass, was included in the plan. The FYP calls for the promotion of a healthy urbanisation plan. It addresses issues such as public transport-oriented urban development, the avoidance of urban sprawl, high-capacity infrastructure and mixed-use development, and the adoption of low-carbon technologies and smart/digital cites (China Daily 2015). Concrete policies and special sectorial programmes will be gradually made by various government agencies in the next couple of years in order to achieve the overall targets.

According to an article published by *Nature* (Tollefson 2016), a joint report from the Grantham Research Centre on Climate Change and the Environment and the Centre for Economics and Policy, China's GHG emissions are likely to reach the peak by 2025, some five years earlier than the target set by President Xi in 2014 (Green and Stern 2015). Such a reduction in energy consumption and carbon emissions is accompanying a new development model—the so-called 'new normal', which embodies a focus on structural changes that can achieve still strong, but lower rates of economic growth of a much better quality in terms of both social equity and environmental sustainability. However, this report also predicts that a continuous increase of energy consumption and carbon emissions from the transport sector would be a focus area for future mitigation. An urban planning model featuring compact, high-density buildings and public transport links should be prioritised (ibid.).

4.3.2 Current Incentives for Eco-Development in China

4.3.2.1 National-Level Incentive Schemes

In January 2013, the State Council published a Green Building Action Plan, which signified that green building development in China has entered a new phase of large-scale development. Then, on 26 April 2013, MoHURD and the Ministry of Finance released the Implementation of Opinions Regarding Accelerating the Promotion of National Green Building Development. In this document, green buildings are recognised as projects that satisfy the requirements from the Green Building Evaluation System (GBES). The Action Plan requires:

- 1 billion metres square construction area of new buildings shall be certified by GBES in 12th Five-Year Plan period (2011–2015);
- 20% of all new construction shall be green buildings by 2015;
- Three types of buildings must comply with green building standards since 2014, including buildings invested by the government such as government offices, schools, hospitals, museums, science museums, stadiums; social welfare housing in provincial capitals and special cities; and large public buildings with single building area over 20,000 square meters such as airport, railway stations, hotels, and shopping malls.

To achieve these goals, there are subsidies provided by the central government to encourage achieving GBES certification. Forty-five RMB per square meter floor area subsidy is provided to new construction that achieves two-star certification, and 80 RMB for three-star certification. There is no subsidy for one-star certification. MoHURD (2012) estimated the average incremented cost are 60, 120 and 300 RMB per square metre for one-star, two-star and three-star certification, respectively, for residential buildings, and 30, 230 and 370 RMB for public buildings, respectively. It is noticed that the current government subsidies cover only a part of the extra costs of green buildings, particularly for public buildings.

The document also calls for the promotion of eco-cities and sustainable urban development. The central government will provide a minimum of 50–80 million RMB to support cities that are recognised as national eco-city demonstration projects. The grant is intended to offset the cost of large-scale eco-developments, such as a new urban area expanded from the existing city. However, the definition and criteria of what constitutes a 'green ecological city' still remain unclear in the document. It is noticed from the current practice of the national eco-city demonstration projects that the visions, strategies and their key performance indicators (KPIs) vary across projects. More discussions will be given in Chap. 5.

Incentivised by the policy mechanisms, in recent years there has been a growing interest in promoting green buildings and eco-cities. At the time of writing, around 280 Chinese cities that have declared an ambition to develop an 'eco-city' or a 'low-carbon city' (China Society for Urban Studies 2012). Green building development is a core requirement for these eco-city projects. For example, Tianjin Eco-City, a joint urban project between China and Singapore, requires the compulsory certification of all buildings within the city through a recognised green building rating system such as GBES, LEED and Green Mark. Such a city-scale implementation has deeply shaped the local construction industry, from design, material manufacturing, construction to operational management, and the green building sector has become a valuable contributor to the local economy.

4.3.2.2 *The Application of Green Building Rating Systems in China*

So far, a number of different green building rating systems have been used in China, including LEED, Green Mark, BREEAM, Green Star and Passive House. Of these, LEED has been the most widely adopted international rating system in China. According to Green Building Map (2015), a website that records certified building projects, there are 195 building projects certified by LEED up to 2014, increased from only 14 in 2008. Inspired by the success of LEED, some other international ratings systems have also been adopted in China. For example, Green Mark from Singapore has been widely used in the case of Sino-Singapore Tianjin Eco-city and BREEAM has been applied to a pilot eco-city project in Changsha (Meixi Lake Eco-City). Similarly, the German Passive House Standard has been used to certify ten buildings in China (up to 2016) since the first passive house was completed in the World Expo in Shanghai in 2010.

Compared to international systems, the Chinese Green Building Evaluation System (GBES) has a much wider application. GBES was officially published by the Ministry of Housing and Urban–Rural Development (MoHURD) in 2006, and was upgraded in 2014. It was developed based on the policy requirements outlined by central government in 2004. These requirements can be summarised as 'Four-Saving and One-Benign', i.e., energy saving, land saving, water saving, and material saving, as well as environmental benign & pollution reduction. To reflect on these policy requirements, GBES provides a three-star rating system for two building types—residential and public. Public buildings in China covers a broad range of structures, including governmental buildings, school buildings and commercial buildings. There are two types of assessment; green building design labelling and green building operation labelling. The latter is granted only after a building has been in operation for one year. Different from some other rating systems which adopt a third-party assessment procedure, the government has been playing a key role in the GBES evaluation. Two assessment bodies—the Centre of Construction Science and Technology and the Chinese Society of Urban Studies, which are both affiliated to MOHURD—are responsible for nationwide certification. Some provinces, and provincial-level cities such as Beijing and Shanghai, can evaluate local green construction projects; however, this only applies to one- and two-star certifications. All three-star applications must be evaluated by the two national assessment bodies.

Akin to other rating systems, GBES comprises eight assessment categories: land saving, energy saving, water saving, material saving, indoor air quality, construction process, operation management and innovation. Within each of them are a number of assessment criteria and benchmarks. Within this context, GBES has become a major player in the market of green building certification in China. It is the first comprehensive system for whole building assessment and has been a national standard since 2006. Before GBES, there were fragmented assessments on various aspects of building environmental performance such as energy, materials and water.

By the end of 2015, there are totally around 4071 projects, or 472 million square metres in floor area, have been certified by GBES according to the 2016 statistics report from the China Society for Urban Studies (2016). Among them, 45% of the total ratings were granted to public buildings, 53% for residential buildings and 2% for industrial buildings. It is noticed that most ratings are for the design stage and only 159 projects have obtained operation labels. This indicates the actual performance of buildings with GBES design labels has not been widely validated in China. It is also noted by Ye et al. (2013) that the distribution of GBES-certified buildings is regionally uneven. More than three-quarters of GBES certified buildings are located in the east coast which is the region with the most rapid economic growth. Although western China accounts for two-thirds of China's territory, its proportion is less than 10%. This reflects the development of green building in a city closely correlates with the economic conditions of that city because a vibrant economy can afford the incremented costs incurred by green buildings.

4.4 LOCAL EFFORTS FOR ECO-DEVELOPMENT IN CHINA

4.4.1 Motivations from Local Governments

It has been observed in recent years that there is a growing interest in promoting eco-development at the level of local government. For example, Beijing has announced that the municipal government would subsidise the energy-efficiency retrofit of buildings, offering 100 RMB per square metre in addition to the subsidy from the central government, and would furthermore subsidise any retrofit which included the installation of solar water heating systems by up to 200 RMB per square metre (Li and Shui 2015). The support for eco-development on the part of local governments is driven by the following motives:

- To reduce energy consumption and related carbon emission—Local governments have been bound to quantitative reduction targets. National targets are broken down to provincial-level targets, and then further decomposed and reallocated to city-, county- and township-level governments.
- To reduce local environmental pollution—Many Chinese cities have been suffering from severe air pollution. As around 78% of the national total electricity is generated by coal burning, consuming energy has been a main reason for local air pollution. To tackle air pollution, local governments in China need to bring down energy consumption in buildings as it is a major consumer of energy;
- To create new business opportunities. The delivery of green buildings, from a view of the entire supply chain, involves many aspects such as planning, design, geological survey, construction, material manufacturing, and facility management. Liu et al. (2012) estimate a potential market of green buildings in China would amount to 1.5 trillion RMB (approx. US $220 billion). This implies an attractive business opportunity for local governments.

To implement the National Green Building Action Plan at local levels, by the end of 2015, 28 provinces, autonomous regions and municipalities have proposed their own timeframes to promote green building construction (China Society for Urban Studies 2016). They cross over the country from prosperous eastern coastal regions to the less developed western regions and some of them have higher requirements than the national one. For example, Beijing requires that all new construction must be certified minimum at the one-star level from 1 June 2013. Jiangsu province sets one-star standards as compulsory requirements for all new construction from 2015 and Chongqing plans to introduce such a scheme from 2020. GBES one-star standards have become compulsory requirements in some cities and incorporated into the administrative procedure for new construction permits.

In addition to direct subsidies, other policy incentives implemented at local levels include:

- Pre-selling commercial properties—Commercial properties with green building certification are allowed to pre-sell depending on the level of the green certification assigned. For example, a certified three-star commercial property can be pre-sold when the foundation is completed; a certified two-star building can be placed in the market when a minimum of one-quarter of the main structure is completed;
- Tax refund for housing buyers to encourage purchasing green buildings, for example, 20% tax will be refunded if buying a two-star green building, and 40% for buying a three-star housing;
- The cost for certification is fully covered by the government, for example, in Shenzhen and Beijing the cost will be covered by the government if a project is certified at the three-star level;
- The mortgage rate can be 1% less if buyers sign contract to buy a green building in Fujian Province;
- Increase of plot ratio for a commercial housing project, for example, an increase of 1%, 2% and 3% plot ration is allowed for projects achieving one-star, two-star and three-star certification in Fujian province.

Local government is a key player in the development of eco-development in China. The use of policy instruments is a significant factor to upscale green eco-development because it can help to create a large market for green projects and cause major changes in construction practice. Local government involvement is one of the essential and effective ways in promoting eco-development through enhancing construction codes, direct financial subsidies and policy incentives. The next section discusses an example of how local government promotes eco-development in China.

4.4.2   Local Initiatives: A Case Study of Ningbo

Ningbo, located on the eastern coast of China and with a population of 7.6 million, has the most vibrant economy in China with a per capita output of US$15,046 in 2013, three times higher than the national average. The city now serves as the economic centre for the southern Yangtze River Delta and has been ranked in the top ten cities for business in China by Forbes.

Ling and Ye (2012) found that green building development in a city correlates strongly with the macro-economic conditions of the city. The annual level of GDP is the most significant individual factor affecting a city's green building development This is because a vibrant economy can absorb the incremental costs incurred by green buildings. It is interesting to note that Ningbo had a GDP rank of 17th among all Chinese cities in 2012 (Elivecity 2015). In contrast, the total green building floor area in Ningbo in the same year was ranked 35th (Ling and Ye 2012), which was behind many Chinese cities that have a lower economic growth.

The city has an official energy saving target of 18.5% for the period of the 12th Five Year Plan (2011–2015), which is higher than the national average of 16%. Building energy saving is recognised by the city government as a focus area to achieve this target. However, green building is still a new concept in Ningbo. Up to the end of 2014, only 19 buildings that have been successfully certified by GBES or LEED, including 9 public building projects and 10 residential projects. Ten of these buildings were certified in 2014. The local government has noted the underdevelopment of green buildings in the city and realised that the lack of a policy instrument is a key factor that impedes the green building development.

Government involvement is believed to be one of the essential and effective ways in promoting green buildings (Chan et al. 2009; Ofori 2006) as it can rectify market failure by introducing market-based incentive schemes (Chan et al. 2009). The reason behind this can be attributed to the fact that green buildings are often perceived as having higher upfront costs for the design, construction and optimisation of building service systems.

To respond to the National Green Building Action Plan, the city has placed this issue on the agenda as a priority, issuing three directives in July and August 2014 to accelerate the development of green buildings in the city, respectively, Opinions on Accelerating Green Building Development in Ningbo, Ningbo Green Building Development Action Plan and Provisions on Adjustment of Conditions for Commodity Housing Presale. The main points of these directives include:

- Set up a Leading Group for Green Building Development chaired by a deputy mayor.
- All government-funded construction must be certified by GBES.
- All commercial buildings with a construction area of more than 20,000 square metres must be certified by GBES.
- Over 10 million square metres of green building will be constructed by the end of 2015.
- Over 1 million square metres of existing buildings will be retrofitted to achieve energy efficiency.
- Commodity housing with green building certification is allowed to pre-sell depending on the level of the green certification assigned: for example, a certified three- star building can be pre-sold when the foundation is completed; a certified two-star building can be placed in the market when minimum one fourth of the main structure is completed.

All quantitative targets have also been broken down to district-level governments by years. The city government further organised a project team in September 2014, including external experts from the municipal construction committee, the University of Nottingham Ningbo Campus and the China Academy of Building Research Shanghai Institute, to examine the effectiveness of the current policy framework and introduce new robust instruments if needed. The project team then conducted a city-wide questionnaire survey to investigate perceptions of various practitioners in Ningbo on the driving and impeding factors for green building development. Between September 2015 and March 2016, the project team has submitted a series of work reports to the city government. All policy recommendations are generally classified into four categories—local financial incentives, local design standards and rating tools, administration, and technologies (Deng et al. 2016). Among the major points to be considered are:

- Local incentives: 15 and 50 RMB subsidies for two- and three-star certification; 70% and 100% exemption of municipal construction matching fee for two- and three-star certification; 20% and 40% refund of property deed tax for two- and three-star certification; 2% and 3% increase of plot ratio for two- and three-star certification;
- Local design standards and rating tools: GBES one-star is compulsory for all new construction and GBES one-star is required for all new construction funded by the local government;
- Administration: set up Ningbo Green Building and Energy Efficiency Commission; set up Ningbo Green Building Experts Pool; encourage and coordinate the third-party consulting service market;
- Technologies: promoting the use of building information modelling approaches in the construction industry; promoting the industrialisation of the construction industry; promoting energy performance contracting; publishing a list of green building products that are suitable for Ningbo.

The city is currently working on the 13th Five-Year Plan and it is assumed that many of the recommendations would be included in the plan and will be implemented by the government in the next couple of years.

As economic development remains a key goal for majority cities in China, strong governmental policy intervention and robust incentive schemes are the main drivers for eco-development, together with multidisciplinary coordination between the involved stakeholders. Furthermore, it is noted that the strategies and practices of promoting eco-development are varying greatly in different cities at different spatial levels. The next three chapters will discuss a number of case studies across the country at the three spatial levels, respectively, with the intention of providing a comprehensive and timely picture of current eco-development in China. These case studies, in contrast with the global examples discussed in Chap. 3, represent the latest policy development and practice of eco-development in China. In each of the case study chapters, we will discuss the case studies based on a defined categorisation at their spatial level. A few of these projects, such as Sino-Singapore Tianjin Eco City and Chongming Eco Island, are internationally renowned. Many are national-level pilot projects, which intend to provide models for future eco-development projects in the country. Most macro-level (city-scale) projects are still under development and are scheduled to be completed around 2020; thus, they also reflect the trend of eco-development in the near future in China.

References

Building Energy Conservation Centre (BECC) (2009) *China Building Energy Conservation Annual Report*. Tsinghua University, China Building Industry Press.

Fan, C.C. (2006) China's Eleventh Five-Year Plan (2006–2010): From "Getting Rich First" to "Common Prosperity", *Eurasian Geography and Economics*, 47, No. 6, pp. 708–723.

Chan, E.H.W., Queena, K., Qian, Q. and Lam, P.T.I. (2009) The market for green building in developed Asian cities—the perspectives of building designers, *Energy Policy* 37, 3061–3070.

Cheshmehzangi, A. (2014) The Urban and Urbanism: China's New Urbanisation and Approaches towards Comprehensive Development, *The International Journal of Interdisciplinary Environmental Studies*, Vol. 8, Issue 3–4, pp. 1–12.

China Society for Urban Studies (2012) China Low Carbon and Ecological Cities Report 2012. China Building Industry Press, Beijing.

China Society for Urban Studies (2016) China Green Building Development Report 2016. China Building Industry Press, Beijing.

Deng, W., Yang, T., Tang, L. and Tang, Y. T. (2016) Barriers and policy recommendations for developing green buildings from local government perspective: a case study of Ningbo China, *Intelligent Buildings International*, pp. 1–17, https://doi.org/10.1080/17508975.2016.1248342.

Green, F. and Stern, N. (2015) *Structural change, better growth, and peak emissions*, the Grantham Research Centre on Climate Change and the Environment and the Centre for Economics and Policy.

Harvey, D. (1985) *The Urbanisation of Capital* (Studies in the history and theory of capitalist Urbanisation), New York: Wiley-Blackwell.

Harvey, D. (2013) *Rebel Cities: From the Right to the City to the Urban Revolution*, London: Verso Books.

Hubacek, K., Guan, D.B., Barrett, J. and Wiedmann, T. (2009) "Environmental Implications of Urbanisation and Lifestyle Change in China: Ecological and Water Footprints." *Journal of Cleaner Production*, 17 (14):1241–1248.

Jacobs, J. (1961) *The Death and Life of Great American Cities*. New York: Random House.

Kristof, Nicholas D. (1993) "The Rise of China." *Foreign Affairs*, 72(5):59–74.

Li, J. and Shui B. (2015) A comprehensive analysis of building energy efficiency policies in China: status quo and development perspective. *Journal of Cleaner Production*, Vol. 90, pp. 326–344, 2015.

Lin, J. Y. and Chen, B. (2011) "Urbanisation and urban–rural inequality in china: a new perspective from the government's development strategy", *Frontiers of Economics in China*, 6(1):1–21.

Lin, J., Zhou, N., Levine, M.D. and Fridley, D., (2008) Taking out 1 billion tons of CO_2: The magic of China's 11th Five Year Plan? *Energy Policy*, 36, 954–970.

Ling S. and Ye, Z. (2012) *Study on the Economics of Green Buildings in China*, China Sustainable Energy Programme, Serial No: G-1110 14964.

Liu, J. Y., Low, S. P. and He, X. (2012) "Green Practices in the Chinese Building Industry: Drivers and Impediments." *Journal of Technology Management in China* 7 (1): 50–63.

Lu, D. (2002) "Rural–Urban Income Disparities: Impact of Growth, Allocative Efficiency and Local Growth Welfare." *China Economic Review*, 13(4):419–429.

Lu, M. and Chen, Z. (2006) "Urbanisation, Urban-Biased Economic Policies and Urban–Rural Inequality." *Chinese Economy*, 39(3):42–63.

McKinsey (2009) *Preparing for China's urban billion*, Report by McKinsey Global Institute, available at: https://www.mckinsey.com/global-themes/urbanisation/preparing-for-chinas-urban-billion, accessed on: 23-Jan-2016.

Nolan, P. and White, G. (1984) "Urban bias, rural bias or state bias? Urban–rural relations in post-revolutionary China." The Journal of Development Studies, Volume 20, 3:52–81.

Ofori, G. (2006) *Attaining sustainability through construction procurement in Singapore*. In CIB W092–Procurement Systems Conference, November 2006.

Smith, N. (2010) Uneven Development: Nature, Capital and the Production of Space. Georgia: The University of Georgia Press.

Tollefson, J. (2016) China's carbon emissions could peak sooner than forecast, Nature News, 21 March 2016, http://www.nature.com/news/china-s-carbon-emissions-could-peak-sooner-than-forecast-1.19597.

Wan, G., Lu, M. and Chen, Z. (2006) "The inequality–growth nexus in the short and long run: Empirical evidence from China." Journal of Comparative Economics, 34:654–667.

Wang, S. and Ouyang, Z. (2008) "The Threshold Effect of China's Urban–Rural Income Inequality on Real Economic Growth." Social Sciences in China, 29 (3):39–53.

Whyte, M.K. (2009) "Paradoxes of China's economic boom." Annual Review of Sociology 35: 371–392.

Wong, L. (1994) "China's Urban Migrants – The Public Policy Challenge." Pacific Affairs 67(3):335–355.

Yang, D. and Zhou, H. (1999) "Rural–urban disparity and sectoral labour allocation in China." The Journal of Development Studies 35(3):105–133.

Yao, Y. (2005) An Empirical Analysis of Financial Development and Urban–Rural Income Gap in China, The Study of Finance and Economics, at Tsinghua Tongfang Knowledge Network Technology.

Ye, L., Cheng, Z., Wang, Q., Lin, W. and Ren. F. (2013) "Overview on Green Building Label in China." Renewable Energy, 53: 220–229.

Ye Q. (2011) Getting off the roller-coaster, UNEP, http://www.unep.org/pdf/op_dec_2011/EN/OP-2011-12-EN-ARTICLE4.pdf.

Ye Q. (2013) Annual review of low carbon development in China: 2010, World Scientific Publishing.

Zhang, D., Aunan, K., Seip, H. M., and Vennemo, H. (2011) The energy intensity target in China's 11th Five-Year Plan period – Local implementation and achievements in Shanxi Province. Energy Policy, 39 (7).

Zhang, K. and Song, S. (2003) "Rural–Urban Migration and Urbanisation in China: Evidence from Time-Series and Cross-Section Analyses." China Economic Review, 14(4):386–400.

4.4.3 Websites

China Daily (2015) Available at: http://www.chinadaily.com.cn/china/2015-12/30/content_22863986.htm.

China Dialogue (2011) Availabel at: https://www.chinadialogue.net/blog/4147-What-s-in-China-s-12th-Five-Year-Plan-/en.

Elivecity (2015) Available at: http://www.elivecity.cn/Index.htmln (Accessed: 15 June 2015).

Green Building Map (2015) (绿色建筑地图) http://www.gbmap.org/.

MoHURD (2012) The Twelfth Five-Year Special Program for Building Energy Conservation; Available at: http://www.gov.cn/zwgk/2012-05/31/content_2149889.htm.

NDRC (2006) The 11th Five-Year Plan: Targets, Paths and Policy Orientation, Available at: http://en.ndrc.gov.cn/newsrelease/200603/t20060323_63813.html.

NDRC (2011) Available at: http://www.gov.cn/jrzg/2011-06/11/content_1881722.htm.

NDRC (2016) Available at: http://www.ndrc.gov.cn/gzdt/201612/t20161202_829076.html.

The State Council (2006) Available at: http://www.gov.cn/gongbao/content/2006/content_443285.htm.

CHAPTER 5

Macro Level: Eco-City Cases in China

5.1 Introduction

In China, eco-city projects are not city-scale projects and they do not cover a whole municipality. They are often developed as new zones, or new districts nearby to or within the municipality of the main urban regions. These can be regarded as experimental projects, satellite towns, or simply developed as new urban expansion zones or new districts to a larger municipality. As argued by Joss (2011), eco-city initiatives were traditionally focused on three primary aspects: the 'physical layout'; 'infrastructural layers'; and 'energy and material flows of cities', all of which shape a tangible spatial form. In China, all three primary aspects are visible, particularly for the larger eco-city projects that lie within close proximity to major urban regions. In comparison to the scale of existing cities in China, these projects are mostly defined as eco-districts, or eco-zones. However, in the national documents and international studies, these projects are referred to as eco-city projects. However, these boundaries are not clear as they represent a wide range of scales under one category of eco-city development. In most cases, they appear as a new district area, such as the case of Guangming eco-city in Shenzhen, Meixi Lake eco-city in Changsha, and so on. In our horizon scanning of eco-city projects in China, we have not come across a complete eco-city development of an existing city or zone by conducting comprehensive retrofitting of the existing built environment. However, there are still comparatively few cases that are focused

© The Author(s) 2018
W. Deng, A. Cheshmehzangi, *Eco-development in China*,
Palgrave Series in Asia and Pacific Studies,
https://doi.org/10.1007/978-981-10-8345-7_5

on improving existing urban environment through effective policy intervention. Having said this, many Chinese cities have a specific agenda with regard to green or eco-development and nearly all of them claim to have specific policies to address these concerns.

For our eco-city categorisation, we explored the existing literature and only came across one categorisation proposed by Joss et al. (2011) who identified three eco-city categories, namely: 'new cities', 'expansion of existing urban areas', and 'urban retrofits'. Based on our horizon scanning of eco-city projects in China, very few cases are focusing on retrofitting the existing urban built environment. However, some cities, such as Guiyang, are exemplary for illustrating how policies moving a city towards eco-development. Nevertheless, we anticipate seeing new cases of retrofitting the urban built environment in the coming years, but perhaps on a smaller scale. Moreover, we moderately argue against such a categorisation of eco-cities, as this can in fact be the categorisation of any new urban development. Therefore, we propose a new categorisation based on the project's stakeholder constellations and the project's status at the national level. Across all tangible cases of eco-city development in China, we have seen a variety of actors in terms of their involvement in the projects. In this regard, we propose a new categorisation that includes three types of eco-city projects in China, namely: (1) eco-city with local initiatives; (2) national flagship projects; and (3) international cooperation projects (also regarded as Sino-foreign projects), which are jointly developed between the local and international partners and often include a strong group of governmental agencies. In this chapter, we explore two eco-city cases for each category. These case studies are selected based on their locations and their current status of development.

5.2 Chongming Eco-Island, Shanghai

Type: Local Initiatives

Project Overview
With an overall land area of 1267 km^2, Chongming Island is the third-largest island in China, and the world's largest river alluvial island. The project sought to develop Chongming Island and two small surrounding islands (Changxing and Hensha) into 'eco-islands'. These thoughts were encapsulated in the so-called Chongming Three Island Master Plan, as part of the earlier 'Chongming County'. Although it is located at a fairly

reasonable distance from the City of Shanghai, Chongming Island only became a district of Shanghai in 2016. Chongming Island and its two surrounding smaller islands are located in the mouth of the Yangtze River, north of Shanghai City. With a total population of around 0.7 million people in 2015, the urbanisation rate of Chongming Island is recorded at 41.7% (Chongming 2017).

The association of Chongming Island with the City of Shanghai is more of a peri-urban interface than an urban area. The overall land area is 20% of Shanghai's land base, and its activity accounted for around 1.2% of Shanghai's GDP in 2012. Since then there are recorded increases of GDP as a result of more connections, business development and real estate projects in recent years. Nevertheless, Chongming Island remains the least developed area of Shanghai City. In 2009, the new Shanghai Yangtze River Tunnel-Bridge was opened to create a better connectivity to Shanghai. The positioning of Chongming Island in relation to Shanghai is changed over the years; i.e., from a satellite town in 1950s and mainly with agricultural farmlands into a key town area of Shanghai, supporting the development of new industries and enterprises producing household electrical appliances, and now into an eco-island (Xie et al. 2017).

Very different to other eco-development projects on the same scale, Chongming eco-island proposes a unique project case in the context of China. This is mainly driven by the geographical feature of this development as an island, and the ambitions of the project. Most people recall the project with the name of Dongtan eco-city, a globally renowned project which received media and research attention several years ago. However, Dongtan was only a small part of this 81 km-long island, lying on the northern side of Shanghai's mega-city. The project of Dongtan aimed to create a new model of eco-city development and was planned to open in time for the Shanghai World Expo in 2010. It was heralded as one of the first Sino-British eco-city projects, involving multiple key actors from both China and the UK, and was intended to develop a new eco-city area in the island, enabling increased population flow and creating a self-sufficient city. The project was first showcased to attract investors in the domestic housing market and involved some of the main real estate companies.

Nevertheless, with some major changes among the local governmental authorities, the project first fell behind the original schedule and then never took place as initially planned. In 2005, contract was first awarded to the British engineering consultancy firm, Arup, and a local developer, the Shanghai Industrial Investment Company (SIIC). The project also involved several Chinese and British state agencies, industries and institutions. The project soon gained an international image and was highlighted as a new eco-city model at the United Nations World Urban Forum, held in Nanjing in November 2008. With zero-carbon development strategies in mind, Arup proposed a comprehensive planning of the eastern corner of Chongming Island. However, the project was halted around 2009, and since then there has been a series of different plans for the overall development of Chongming Island.

Although some argue that there are separate directions of development between Dongtan eco-city and Chongming eco-island (Chang and Sheppard 2013), it is now clear that the Dongtan area is now only a major part of the island's eco-development strategy. With SIIC's continuous involvement, Dongtan is now home to a new wetland park, which preserves the existing ecology and minimises construction in the area. Nowadays, Chongming eco-island is a distinct national-level eco-island project, with current planning in progress towards achieving a world-class standard. It is also formally included in the 13th Five-Year Plan for National Economic and Social Development in Shanghai (Shanghai Municipal Government 2016). Although it is considered to be a national-level pilot project, its local initiatives are worthy of attention, including several community-led eco-farm projects and 'Yingdong Ecological Village', both of which represent the local citizens of Chongming Island (Xie et al. 2017). Almost in parallel, the island includes a major international experimental eco-community development of 'Chenjia Town', which is strategically located nearby the entry point to the island (ibid.). In the later sections, we elaborate on these key eco development projects.

Project Layout and Planning

In 2006 the Shanghai Municipal Government and Chongming County government issued the Chongming Three Island Master Plan, covering the rest of Chongming County outside Dongtan eco-city (Fig. 5.1). This was a locally-driven independent plan based on Arup's proposal, focusing on smaller-scale environmental improvements and aiming to develop Chongming Island and two small surrounding islands (Changxing and Hensha) into 'eco-islands'.

Land on all three of the islands is zoned into several functional regions, including ecological system demonstration areas, leisure and tourism, sport and vacation, a garden city, education and innovation, forest, theme park, conference centre and offices, and a shipbuilding industry special area (Shanghai Municipal Government 2016). These are set out as follows:

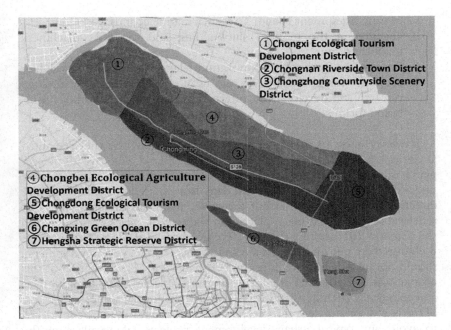

Fig. 5.1 Overview of Chongming Eco-island plan (adapted and redrawn by the authors based on the most recent masterplan strategy development of Chongming Eco-island provided by the local government in the project exhibition hall)

1. Forest Garden Island: To form an ecological-conservation island mainly based on the Yangtze River Estuary Wetland Reserve, the International Migratory Bird Reserve, the Plain Forest and the River Estuary Water System.
2. Ecological Inhabitation Island: To form an ecological inhabitation island that is properly designed, quietly surrounded, conveniently located and culturally advanced.
3. Leisure Vacation Island: To form an ecological tourist island that mainly caters for the leisure vacation, sports and entertainment, recuperation, training, conferences and exhibitions.
4. Green Food Island: To form an eco-farming island featuring the organic agricultural products industry, characteristic cultivation and aquaculture industry and the green food processing industry.

5. Marine Equipment Island: To form a marine economy island mainly engaged in the modern shipbuilding and the port machinery manufacturing.
6. Science and Technology Innovation Island: To form a knowledge-economy island assigned to the headquarters offices, technology research and development, international education and the consultation forums.

However, most of the initial ideas never happened and they remained as options for the future development of the island. The overall project plan has been altered on several occasions and few small-scale projects are under development as at the current date. It is still unclear what the future direction will be, as the current projects remain incomplete. The most remarkable change has taken place after the visit of mayor of Shanghai to the island in 2016, where the new developments of medium- to high-rise buildings were put on hold. As a result of this change, new developments should not include any buildings higher than 18 m. The intention of this change was to bring back the image and spatial feeling of Chongming as an ecologically-friendly island, capturing the ecological values and image of the island. In our view, this means a new direction should be considered to avoid any typical urban development projects taking place in the Chongming region.

According to Xie et al. (2017), the overall plan for Chongming has been in transition since its inception. The following three stages summarise the progress since 2005:

- In 2005, "Overall Plan for Chongming Three Islands (2005–2020)"
  Planning ideal: (1) adhere to *"Ecological Modernisation"* pathway; (2) Conserve adequate "natural ecological space" and "space for future international projects planning"; (3) *"Three Concentration"*: concentrating land operation, concentrating industries in parks, concentrating farmers' residence in towns/cities.
- In 2010, "Chongming Eco-Island Construction Outline (2010–2020)"
  Action programs: focusing on ecological and environmental indicators, taking into account economic and social indicators. Established the evaluation index system for 2020 Chongming eco-island construction, and developed action measures.

- In 2016, "Thirteenth Five-Year Plan for National Economic and Social Development of Chongming (2016–2020)
   Implementing "*Eco+*" development strategy, and introducing more demanding indicators.

In this respect, what we can witness is a clear case of eco-development in transition. With several major changes in the past, the project is now under progress with clear eco-strategies. The approach to the overall island development is very much centralised with several county and town developments. The current phase of the project only demonstrates few of these new projects that are currently under construction in various parts of the island. In what follows, we highlight some of the main eco-development projects at various scales and models in several parts of Chongming Island:

Individual Eco-Farm Projects—With no governmental policy or incentives, the eco-farm projects are mainly led by the individuals of the local community. Some examples of these bottom-up initiatives are located in Chenxi Collage of Chenjia Town, close to the entry point of the island, and in Meiyuan village. The main intention of such small-scale projects are to create the idea of new eco-farm industries and promoting the existing rural community (some information is based on interview transcripts contained in Xie 2016). Such projects are believed to be key to local industries and the enhancement of local communities, where the least impact on the natural environments is expected. The location selection is based on the three key factors of: (1) low rental prices; (2) environmental quality; and (3) proximity to the Shanghai region (ibid.). It is unclear if such small-scale initiatives would be sustainable for a long period of time, but current progress show some success for community-led movements in several parts of the island.

Yingdong Ecological Village—Initiated by the local village committee and collective initiatives from the community, Yingdong Ecological Village represents one of the key bottom-up and local projects of Chongming Island. In 2013, after the reform of property rights system, a collective-owned enterprise in Yingdong Village was established. The village community became the stakeholders, enabling them to work as part of their new vocational enterprise. The project had the central goal of 'ecological development' and included a building of a 10 km-long ecological river. Led by Tongji University and the Shanghai Research Institute of Building Sciences Group, the project was to create a new living environment for the local communities (some information is based on interview transcripts in Xie 2016).

Chenjia Town—In close proximity to the Dongtan area lies a major new eco-development of Chenjia Town. Proposed to have new water bodies and a lake, the new development is one of the key eco-development projects of Chongming Island (Xie et al. 2017) and, since its inception, has gained major attention in the media. Another similar type of project is Chengqiao New Town, which is around 38.7 km² in total. The new development of Chenjia Town is roughly 2.5 times larger, at 94 km². This large-scale development is three times the size of the Sino-Singaporean Tianjin Eco-city, and is the major business hub of Chongming Island. Its zoning plan is part of the island's county and town development strategy, including ten function zones, such as the International Experimental Eco-community, the International Forum Business District, Yu'an Modern Community, and the Dongtan International Education R&D Area that constitutes the 'Central Town Area'; while the other six areas, including the Wetland Tourism Area, the Riverside Leisure Sports Area, the Green Industry Area, the Theme Park Area, and so on. The new development proposes for urban fringe development and the low-carbon industrial zone of the island. The existing Dongtan wetland park is enclosed in the new area. The project layout is an ecological city in the island garden style, and focuses on ecological preservation. The current strategy is believed to be capable of being scaled up for other parts of the island, where the proposals are focused on additional green zones and green spaces, population growth and limitations on the new built environments.

Green Technologies
Based on the Chongming Eco-Island Development Outline 2010–2020, a number of green features are proposed. The proposals include a long-term development plan for green industries (based on Shanghai Development and Reform Commission 2010), which has so far catalysed the completion of several tourist farms, vacation homes, forest and wetland parks, a conference centre with a five-star hotel, and trail routes. In 2009 and 2010, parts of some agricultural villages at the eastern end of the island were relocated to create space for modern medium- to high-rise housing zones. As mentioned earlier, the building heights are readjusted to reflect the local context and the Chinese identity. Moreover, the opening of the Tunnel Bridge connecting Shanghai and Chongming Island opened in 2010 further enhanced the eco-island fortunes.

With regard to waste management, Chongming Island is actively promoting the application of solid waste classification and source reduction in a number of different ways. Confined transition and transportation systems have effectively avoided secondary pollution during the transition. The Chongming Solid Waste Disposal Site was rated as a Grade I hazardous-free disposal site. A kitchen waste treatment plant has also been built and put into operation. The Chongming Solid Waste Disposal Utilisation Centre has been launched to further promote resource utilisation. The recycling and reuse of both agricultural and construction waste is also steadily increasing in the Island. In addition, a low-carbon energy structure mainly supported by clean electricity and supplemented by renewable energy and a smart grid. The other additions include a low-carbon industrial structure framed by advanced eco-agriculture and service industries, a low-carbon infrastructure system through the localisation and application of green building designs and low-emission vehicle technologies, a natural carbon sinking system supported by forests and wetlands, and a relatively low-carbon lifestyle anchored on the Chinese traditional philosophy of unity of nature and human-being.

In 2011, UNEP and the Science and Technology Commission of Shanghai Municipality (STCSM) came to an agreement to involve UNEP in evaluating the development plans and progress of Chongming eco-island. UNEP's focus has been mainly on suggesting methods of ecosystem management and sustainable development practices that are applicable to the context of Chongming (UNEP 2014). After the first round of evaluations, UNEP provided specific action plans in seven key areas: (1) Society, life and culture; (2) Biodiversity and protected area; (3) Water management and conservation; (4) Solid waste management; (5) Low-carbon economy and energy efficiency; (6) Agriculture and organic products; and (7) Transportation (UNEP 2014). Under each key area, specific action plans were provided for the next phase of development. Additionally, UNEP (2014, p. 8) provided four general recommendations for Chongming Eco-Island, including key action points to:

1. Establish the Chongming Special Ecological Zone for Eco-Civilisation;
2. Accelerate the dissemination of transformative scientific achievements through a systematic promotion of demonstration projects to amplify the role of science, technology and innovation (STI);
3. Strengthen and expand the Chongming Eco-Island International Network as a platform for exchanging best practices, lessons and promoting the Chongming Eco-Island Construction Model;
4. Strengthen human resource capacity, build support systems and enhance stakeholder coordination and collaboration.

Currently, Chongming follows multiple eco-strategies that consist of various green technologies and design. Since its development in 2016, the key direction is focused on an 'Eco+' development strategy, enabling the promotion of ecological economy, development of modern agriculture, building new outdoor activities and preserving the ecological areas of the island. By doing so, the project aims to implement a comprehensive 'pollution source control', by continuing the closure of heavily-polluting industries and factories. For instance, over the past 15 years, about 1000 factories in the Chongming region have been closed down (Xie et al. 2017). The other major action plan is to have more tree planting, in order to increase the forest coverage. This increase is a continuation from previous attempts, from 10% forest coverage in 2000, to 22.53% in 2015. 'Sewage treatment' is also considered part of the overall plan, where the previous attempts have increased the treatment of the county and town areas from 10% in 2000 to 85% in 2015 (ibid.). In additional to this key progress in the past, the project has achieved the most optimal air quality in the region of Shanghai municipality, achieving the lowest rate of PM 2.5 amongst all districts. Not only does it supply the core water source of the whole city of Shanghai, but it owns the best water quality as part of the water management and reservoir development. These key factors demonstrate the island's strong focus on ecological development.

In contrast to other eco-cities, the Chongming development was inspired on the basis of the 'failed' Dongtan eco-city project. Having said this, some of the smaller-scale projects were undergoing before the proposal of Dongtan eco-city. Currently, most of Chongming Eco-island's key development projects are still in their infancy and some show tangible progress regarding the preservation and enhancement of the existing ecological zones. It is unclear when the project is set to be completed; as it stands at the moment, however, some of the smaller-scale projects need further attention. And if they are to be considered, propositions can be made to scale up some of the key local initiatives. However, Chongming eco-island represents a fusion between local initiatives, national programmes and international cooperation. The focus here is on local initiatives, since the project has gained significant attention from the local authorities in recent years. Also, as expressed by Chang and Sheppard (2013, p. 71), Chongming eco-island is more of a "'local version of sustainability … overlaid onto Euro-American conceptions of eco-cities, one that is constructed and conditioned on, and shaped by, local desires for economic development, the geographical imagination of an island, and the ambition to make global cities'. This factor is also one of the reasons why most development plans were not rushed for construction, and in comparison to many other cases of eco-development in China, much better care has been given in various stages of the project's development.

Case Study Reflections
It is noted that main reason for the 'green development' of Dongtan and Chongming was a desire for environmental and economic development in the light of strengthening relationships with Shanghai (Chang and Sheppard 2013). Thus, it can be seen that Eco-cities and/or Eco-islands can also be a form of green capitalism, with the island's natural capital being the only resource to enable Chongming Island's economic development and realisation of overall sustainability. Also, according to UNEP's report (2014), and after several years of change and development, we can witness a remarkable number of development projects and transitions; however, much more needs to be done to achieve the maturity status so envisaged. As suggested by UNEP, exact data on technologies deployed as compared to the baseline will also need to be monitored closely.

5.3 Guiyang City, Guizhou

Type: Local Initiatives

Project Overview
Located in South-western China and east of the Yunnan–Guizhou Plateau, Guiyang is the capital city of Guizhou province with a population of 4.7 million by the end of 2014. Guizhou is a less developed region in China, ranking 25th in terms of GDP among 31 provinces and province-level municipalities in 2014. Likewise, Guiyang was ranked 28th in GDP among all provincial capitals in 2014. However, both Guizhou and its capital city have gained momentum in recent years with a higher growth rate than the national average; for example, in the first half of 2017, the city Guiyang has the highest GDP growth rate of 11.6%.

Guiyang is also China's top-ranked city in terms of implementing ecological civilisation. In December 2012, the city was approved by the central government to become the country's first national model city for ecological civilisation. It is also a national pilot city for urban innovation development, and a leading area for coordinated urban and rural development. Guiyang is the first city in China to propose and implement local regulations on the development of circular economy and ecological civilisation. It provides an example of promoting sustainable development through strong policy interventions.

Project Layout and Planning
Ecological civilisation is a new term, proposed by the party in the 18th National Congress in November 2012. In the keynote speech delivered by President Xi in the congress, he stressed that building a beautiful China is an important part of the Chinese dream of national rejuvenation. Xi emphasised four tasks to promote ecological civilisation—optimize land comprehensive exploitation; comprehensively promote resource conservation; reinforce the efforts to protect the ecological environment; enhance institutional systems for ecological civilisation.

One month after the national congress, in December 2012, Guiyang was approved as the first city by the central government to construct a model city for ecological civilisation. Guiyang was picked out largely because of its long history of maintaining a sustainable profile during the course of rapid industrialisation in China. A chronological record of the awards for its sustainability performance in recent years are listed below:

- The first national pilot city for circular economy in 2002;
- Guiyang Master Plan for Constructing Circular Economy Eco City was launched in 2003, this is the first urban masterplan focusing on circular economy;
- The first global pilot city for circular economy in 2004 by UNEP;
- The first batch of the national pilot city for low-carbon development in 2010;
- The first national pilot city for constructing eco-civilisation in 2012;

As the most renowned city in terms of developing circular economy and eco-civilisation in China, it is the country's intention that the practices pursued through the pilot projects in Guiyang should be applied to other Chinese cities, in particular those cities that are less developed and facing twofold challenges for improving economic growth and the urbanisation level while moving towards sustainability.

In 2013, over 400 foreign guests from foreign government agencies, heads of international organisations, internationally renowned scholars and heads of think-tanks, and leaders of international eco-environmental protection enterprises attended the first International Conference on

Ecological Civilisation in Guiyang, which has become an annual event since then, and is the only state-level and most important international forum on green development and ecological civilisation in the country. The Guiyang Consensus 2013 was issued by the participants that addressed four pillars of eco-civilisation—to accelerate green development and green industrial transformation; to promote social harmony and inclusive development; to take the strictest measures possible for the repair of damaged ecosystems and depleted natural resources; and to popularize ecological values across the whole society (IUCN Environmental Law News 2013). A specific theme is selected for an annual conference. These themes have responded to the recent policy upgrading and enriched the concept of ecological civilisation. The themes selected from 2013 onwards are listed below:

- 2013 theme: Green reform and transition—sustainable development led by green industry, green urban development and green consumption;
- 2014 theme: Driven by policy reform and global partnerships—governments, enterprises and public: policy framework and roadmap for green development;
- 2015 theme: Entering a new era of ecological civilisation—new agenda, 'new normal' and new action;
- 2016 theme: Toward a new era of ecological civilisation—unity of knowledge and practice;
- 2017 theme: Sharing the benefits of green development—green development and poverty relief.

In 2012, the Guiyang Plan for Building National Eco-civilisation Model City was approved by the central government. This indicated that Guiyang had advanced the concept of eco-civilisation to guide its economic and social development and that it is determined to become the eco-civilisation pioneer in China. In 2015, the local party commission proposed the targets for the period of the 13th Five-Year Plan (2016–2020):

- Building an innovation-oriented city;
- Develop a comprehensive innovation zone for Big Data technology;
- Building an eco-civilisation model city, and;
- Building an overall well-off society at a higher level.

In March 2016, the Guiyang 13th Five-Year Plan for Building Eco-civilisation has been approved by the Municipal People's Congress, which proposed a range of targets to be achieved by 2020. Figure 5.2 indicates the roadmap and the targets to be achieved by 2020.

Setting Targets for Achieving Comprehensive Eco-Civilisation in 2020

The municipal government proposed a set of indicators in 2015 to monitor the progress towards eco-civilisation. This comprises 33 controls targets and 37 optional targets which are classified into six categories—ecological economy; natural resources conservation and intensive use; improvement of ecological environment; liveable urban and rural environment; ecological culture; and ecological civilisation system. As the first city aiming to achieve eco-civilisation, these targets represent what an eco-civilised city should be like at the current economic and social stage in

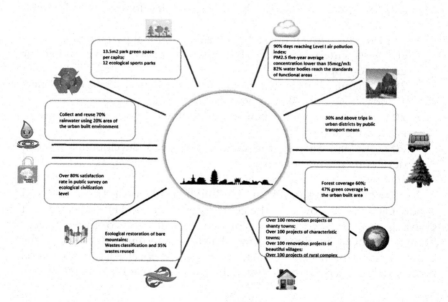

Fig. 5.2 Roadmap and targets to be achieved by 2020 for constructing ecological civilisation in Guiyang. Source: Adapted and redrawn from Guiyang Evening News, http://wb.gywb.cn/epaper/gywb/html/2017-10/01/content_32894.htm

Guiyang, and create a model for other Chinese cities. Apart from a number of ecological targets, such as energy and water consumption, renewable utilisation, promoting green buildings and urban air pollution, it is noted that the plan also incorporates indicators in the socio-economic and institutional arenas, for example, health insurance coverage, urban unemployment rate. These targets are broken down and allocated to governments at different levels such as districts and townships.

Restructuring of Industrial Pattern
The city has moved its industrial pattern to be increasingly service-economy-oriented by zoning the territory into four types of land use:

- Exploitation-prohibited area;
- Exploitation-constraint area;
- Exploitation-optimised area, and
- Exploitation-implemented area.

Project proposals have been evaluated carefully based on the degree of environmental friendliness. One priority area chosen by the government is Big Data industry—which has been designated as one of the three strategic sectors in Guizhou province. Accordingly, Guiyang has released preferential policies to boost Big Data industry. Guiyang aims to host several international and national data centres. Renowned international companies like Hewlett-Packard and the national cloud computing have launched their data centres in the city. The city also enhances the construction of infrastructure facilities for IT technologies. For example, the city is working to provide free city-wide Wi-Fi. The project's first phase was completed in May 2015 to allow free Wi-Fi on the city's major roads and main public areas, covering an area of 12.8 square kilometres.

Promoting Low-Carbon Living Community
It is an important part of Guiyang's strategies to encourage changes to low-carbon living behaviours and consumption patterns. The city has implemented a low-carbon community scheme and established a set of indicators that provided a quantitative way to monitor low-carbon community development; for example, an estimate of carbon emissions of 161 tonnes a year has been saved by replacing the street lights and installing smart control lighting system in households in a small residential district. The government also publishes regularly green product lists, including

energy-rated, water-rated and environmentally- labelled products to encourage the change of consumption patterns. In particular, 80.5% of government purchases must be from green product lists by 2020 that is a control target for eco-civilisation construction.

Promoting New Technologies in Sustainable City Management
A so-called 'ecological cloud' platform is currently under development, which will use the city's booming Big Data technologies and cloud platform to promote sustainable city management. Guiyang aims to establish an ecological cloud computing database in 2017 at three levels—city, county and township and in 2018 incorporate forest management, urban greening and environmental information databases to form an integrated 'ecological cloud' platform.

Case Study Reflections
Unlike Beijing and Shanghai, Guiyang is a less developed city and faces twofold challenges—economic growth and sustainable development. Some cities undertake industrial transfer by accepting industries with high energy and resource intensity from prosperous regions. This is an easy choice and a lure. Guiyang, however, has devoted itself to restructuring its economic system, adopting new technologies to promote the idea of a sustainable city, encourage low-carbon consumption pattern and, most importantly, develop a vision and measurable targets to monitor its progress towards eco-civilisation.

5.4 Meixi Lake Eco-City, Changsha, Hunan

Type: National Flagship Project

Project Overview
The Meixi Lake Eco-City (MLEC) is one of the first batch of eight state-level eco-city model projects in China (Franshion Properties (China) Limited 2014; Cheshmehzangi and Deng 2015). The total cost of this large-scale project is estimated at £6 billion (equivalent to approximately US8 billion), including almost 15 million square metres of new development. The project was first initiated in February 2009, after the bilateral agreement was made between the Municipal Government of Changsha and the real estate developer Gale International. The first intention was to develop a new green development of 1675 acres (or 1887 for the larger area, in other sources), as a new satellite town to the city of Changsha, the capital of Hunan Province. The developer team previously worked on the

Songdo International Business district of Incheon City, in South Korea (Joss et al. 2011). The project is currently partially completed and is partly operational. It is due to be completed by 2020.

The backbone of the MLEC project is focused on three key aspects of 'plan, develop, and operate', central to the new development strategy towards an ecological city (Gale International 2010). Some of the earlier works focused on project marketing and a feasibility analysis for the next steps of developing and implementing the final masterplan and infrastructure design.

With regard to the project's governance and financing, apart from the support of national and municipal governments, the project benefited from the financial support of China Merchants Bank (CMB) with an additional 10 billion RMB in credits for key new development areas and enterprises of the first phase of project (Hunan Government 2009). The project also benefits from an extensive partnership network of international planning and engineering firms, and real estate developers, such as Kohn Pederson Fox (KPF), Arup, Gale International, Atkins, and Cisco. In the project development plan, it is stated that Cisco plans to deploy video-networking technology and energy- management software tools city-wide and meld municipal systems, such as education, health care, transportation and hospitality into a common network (Woyke 2009). In this regard, Cisco focuses on the provision of several services as part of their Smart + Connected communities initiative. In summary, the project is supported by multiple partnerships, but is principally regarded as one of the national eco-city flagship projects.

The new eco-development was first named the 'Meixi Lake District' of Changsha. Its title was later altered to the 'Meixi Lake International District'. The project is located in the core region of Xiangjiang New District, located in the outskirts of Changsha and around six kilometres from the centre of the city. Once completed, it is anticipated that MLEC house around 180,000 residents (Alusi et al. 2010). According to the project's masterplanning and design team, KPF, the project is proposed to offer 'a new model for the future of the Chinese city', as it focuses on merging the metropolitan features and natural environments as exist around the new development area. The MLEC also proposes for the construction of a smart grid system, new urban agricultural areas, innovative transport networks, and waste energy recovery (McGraw Hill Construction 2010). The new development also fits in well with Changsha's expansion plan and economic development strategies.

Project Layout and Planning

Benefitting from good feng shui, the MLEC development is backed by Taohua Ridge and the Yuelu Mountain Range, and is surrounded by the centrally located Meixi Lake with a total surface area of 18.05 km². There is a long islet inside the lake, with several smaller islets are part of the park/green space development of the MLEC. The plan is that once the project is completed, it should appear as a national eco-city model, and propose for new benchmarks and standards for future developments (Franshion Properties (China) Limited 2014). This was also highlighted as part of Changsha City's 12th Five-Year Plan (FYP), in which the project was proposed in two phases of development.

Similar to the layout of an earlier and similar project, Songdo District in South Korea, the masterplan is comprised of a high-rise business district located around Meixi Lake and is linked to several residential communities by a series of canals (Joss et al. 2011). The project benefits from linking to Changsha's metro system, bus system and also its own boat system. Most of the area is under development for grey and black water treatment systems and distributed energy plants (ibid.). In addition, the MLEC features 45 ecological indicators ordered into eight key categories which cover the overall planning and design of the project. These broad categories are: (1) Green culture; (2) Urban planning; (3) Ecological environment; (4) Energy; (5) Water resources; (6) Solid waste; (7) Transport; and (8) Green building (Cheshmehzangi and Deng 2015). For instance, all newly-constructed buildings should be certified as green buildings, and the UK's Building Research Establishment (BRE) has played a significant role in achieving this target plan. All green construction projects are also subject to constant monitoring, which is conducted by the energy consumption system throughout the whole design and construction processes (Franshion Properties (China) Limited 2014). Since 2012, the project has also been also named as one of the nation's most representative in terms of resource-saving and environmentally-friendly communities that has enhanced the national positioning of the project.

The overall masterplan of the MLEC project includes nine core areas as functional areas of the new Meixi Lake Region (Fig. 5.3). For instance, the Central Business District (CBD) project is the central part of the first phase of the Changsha Meixi Lake Primary Land Development Project (Franshion Properties (China) Limited 2014). It is important to note that the new development model proposed by the masterplanning team is now known as the 'Meixi Lake Model' in the Chinese industry. In this

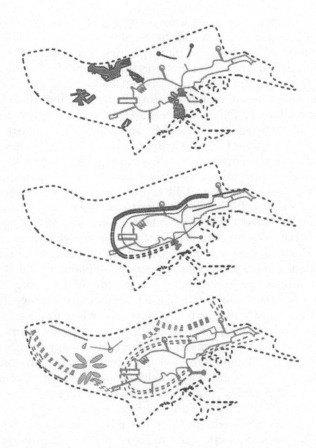

Fig. 5.3 The multiple layers of green strategy for green spaces, green corridor, and key spatial qualities of Meixi Lake Eco-City (adapted and redrawn by the Authors, from the KPF masterplan documents)

particular model, the approach is to integrate the new technologies and the new proposed planning methods to develop well-connected and ecologically-sound living urban environments. Unlike other eco-city cases in China, the MLEC model is planned in a radial layout with the combination of residential neighbourhoods and villages in the surrounding areas. In this respect, the project has the central primary design element of 'water', with multi-functions, from ecological values to leisure activities. It is also regarded as a central park of the new eco-city providing a better natural cooling effect, better natural qualities, better spaces for various uses and events, and partly operational for transportation needs (e.g., boats and ferries). The radial layout also enables a more manageable layout for neighbourhood design creating eight neighbourhood clusters that accommodate 10,000 people each (KPF website). The connected transportation network of the MLEC, and in connection with the surrounding areas and the City of Changsha, is considered a very efficient model that aims to reduce needs for car usage and target low-carbon modes of local transportation. The radial layout, according to the masterplanning team (obtained from KPF), also offers a highly efficient transportation system, a well-established infrastructure for water management, and a well-distributed network of energy plants and urban agriculture across the development. The project also benefits from holistic design strategies which are seen as some successful global examples, and focus on advanced environmental engineering, pedestrian planning, cluster zoning and garden integration in the built environment (KPF website).

In addition, Atkins Global (2013), a UK-based consultancy firm, provided a comprehensive low-carbon plan for the entire development. The other technologies included in the project were the provision of water-cooled air-conditioning systems for buildings and energy-saving elevators (Alusi et al. 2010). The intention was to cut energy consumption by 20% for the main services of individual buildings across the development. In addition to energy-saving strategies, new energy technologies, such as digital signs and 'stick-on film' for building façades, were introduced to the newly constructed buildings. Cisco has also played a significant role in implementing some of their smart technologies and innovations in the project.

Case Study Reflections
It is repeatedly reported that construction of the MLEC project is on schedule (Franshion Properties (China) Limited 2014) and the project has entered its second phase of development. There are no official data available on building occupancy, but based on our observation on several site visits (2014 and 2015), the real estate projects are partially occupied and mostly completed as part of the project's first phase. The project is positioned to develop a new international image of Changsha City and a new model of eco-development in Central China, which embraces the

concept of a two-type model district, i.e., achieving both and environmentally friendly development and a resource-saving new city project. The WLEC is believed to adhere to the group's initial strategic thinking by providing an ecological development as well as achieving a high-end development with quality.

Once just a farmland in a traditional rural setting, the MLEC is now one of the remarkable examples of new urban developments in China. As one of the pioneering national eco-city models in China, the MLEC project represents a unique model of itself. The project, while named 'international' and involving several key international actors (mainly in planning, feasibility analysis, technologies and engineering), is a national-level eco-development project. The low-carbon and green approach to building design is consistent across the whole region, and the project represents a model of multiscalar and multifunctional planning.

While we can question the ecological aspect of the project since the new built environment is built over the existing agricultural lands, it is significant that the MLEC has undertaken a comprehensive plan to minimise the overall impact from the new built environment on the existing area. Hence the project's low-carbon agenda appears stronger than some of the other eco-city projects of the same scale in China. The MLEC, in many ways, is argued as not only a new city development, but is also comprised of new urban planning methods that makes it into a national model.

5.5 Guangming Eco-City, Shenzhen, Guangdong

Type: National Flagship Project

Project Overview

Guanming Eco-City (GEC) is one of the first batch of eight national eco-city projects in China in 2012. It is often described as 'Shenzhen Guangming New District' and is now one of the seven districts of the City of Shenzhen in South China. The Guangming region used to be part of the larger Bao'an District of Shenzhen, but was made into a new district in 2007; after which the proposal of GEC was under consideration. The new development project has a combined approach to eco-development and low-carbon construction, including key strategies for optimising the existing structure of urban space, enhancing green municipal planning, developing low-carbon industrial development, forming a green transport system, and promoting green building design. Since the establishment of the new district in 2007, rapid development has taken place across the district, one of which is the GEC project. The new development is aimed to scale up at a later stage, while currently creating a hybrid of eco-city and low-carbon town model. The GEC project has so far received three national recognitions for green development, including: National Green Building Model Town in 2008; China's first low-impact development model town in storm water management in 2011; and National Green and Ecological Model District in 2013. The project is also highlighted as one of the eight national-level pilot projects for reform demonstration.

In recent years, GEC has become one of the national projects that has received significant international attention (Yu 2014) and is led by several layers of governmental bodies, many high profile at the national level (Ji et al. 2017). This new eco-development project also acts as a sub-centre to the City of Shenzhen, where the new high-tech zone is designated as an industrial hub to support surrounding industrial estates and high-tech firms (Zacharias and Tang 2010). The construction of GEC was initiated in 2011, after the official announcement of national demonstration projects were made by the National Development and Reform Commission (NDRC) in 2010. The project follows the 'Shenzhen Guangming New District Green City Index System', with a medium- and long-term plan for Shenzhen's low-carbon development. The overall plan (2011–2020) is drafted by the Shenzhen Development and Reform Commission.

In 2007, a merger between two street districts, 'Guangming' and 'Gongming', created the Guangming New Town District, which used to be a state-owned farm. It is located to the northwest of Shenzhen, and back then as one of the rural outskirts of the city. Its location makes it the gateway linking Guangzhou–Dongguan–Shenzhen–Hong Kong. With such a strategic location, in the middle of the 'urban and high-tech industrial development axis, Guangming has rapidly developed and urbanised over the years, while preserving some of the surrounding mountainous landscape, the Maozhou River, nearby water reservoirs and lakes. Since then, Guangming has developed both physically (e.g., infrastructure and the built environment) and economically. The GEC project is in reality part of a larger economic growth zone in the city of Shenzhen.

The total area of the Guangming district is around 156.1 km^2, and nearly 46% of the area is given to urbanisation and new urban expansion projects, including the development of GEC as part of the overall planning of the new district. The remaining area is officially part of the Basic Ecological Control Line, which also acts as one of buffer zones to the expansion of Shenzhen. Although not very urbanised, the whole district is home to about one million people (some sources indicate 0.8 million people), largely part of the existing farmland and new high-tech industries.

Project Layout and Planning

The eco-city indicators for this eco-city was developed in close collaboration between the Shenzhen Branch of China Academy of Urban Planning and Design, the Chinese Society for Urban Studies, Cardiff University and the Shenzhen Municipal Institute of Urban Planning and Design (Yu 2014). A close collaboration was also established with close collaboration with environmental consultants Fulcrum and leading engineering firms Techniker, and Alan Baxter Associates.

The total size of the central part of the GEC project is around 7.4 km^2, with the defined areas for development within the new district covering 3.4 km^2 (340 ha), of which 67 ha are the built-up areas. The land uses include industry, rural housing, administration and office plots, road, infrastructure and public utilities (Fig. 5.4). The Guangming area is essentially a fertile farming area, thus the planning sought to integrate the farming into eco-city planning. As mentioned in their policy brief, Birch et al. (2014) argue that the new district is planned to be a 'Green New City, Entrepreneurship New City, and Harmonious New City', following the two key national directions of the new urbanisation plan and industrialisation. Alongside development of new services and a new industrial hub for Shenzhen, Guangming aims to develop high- quality living areas, strong public facilities and tourism industry. In this respect, the new development is divided into three zones of: (1) traditional industry development zone; (2) high-tech industry development zone; and (3) ecological industry development zone. The latter is where the GEC project is located, and includes agricultural activities and eco-friendly tourism. The key spatial strategies that are considered as part of the planning of the new eco-city project include (ibid., p. 33):

- A convenient network of public transport and a slow-traffic system;
- Compact and transport-oriented development (TOD) in station areas;
- Introducing greenery from the ecological zone and surroundings by green wedges;
- The use of recycled energies, waste, and rainwater;
- Green building technologies.

Fig. 5.4 Overview of Guangming Eco-City plan (adapted and redrawn by the authors based on the most recent masterplan strategy development of Guangming Eco-city)

The mentioned green building technologies explicitly focus on the use of measures for the emission reductions, vertical farming methods and using the roof top farms of 12 circular towers, land allocation for solar panels at the ground level, and water reuse of up to 50%.

As part of the whole district development, there are more than 40 green projects that are proposed as part of the ecological and low-carbon development strategies. All aspects are incorporated in the local development, whereby all new projects are anticipated to address the green target plan of Guangming new district.

As described in their report, Wang (2012) highlight the key green development characteristics of GEC (2007–2020) that are adapted in planning process of the project, as shown in the following: (1) TOD mode plays a major role in increasing land-use efficiency and preserving the ecological values of Guangming, creating a better growth boundary management and compact layout; (2) the overall plan encourages a better distribution of large-scale infrastructure that attracts low-carbon and highly efficient economic activities in the local setting; (3) pollution control is considered holistically as part of the project plan, including sponge city development, urban flood control, and development of new urban ecological parks to protect the water system associated with river branches of Maozhou River; forming an ecological framework that reflects on green space expansion and green land preservation; (4) the integration of green transport plan of the Guangming into the city's transit network, including key public transport systems and other public services in the district; (5) consideration of integrated design for new road network and public transport system, allowing for better arrangement of transportation layout, non-motorised traffic priority zones and express traffic corridors for better connectivity to the surrounding regions and the central Shenzhen; and (6) promoting sustainable methods for engineering projects allowing for sustainable design, using local products and achieving high performance after completion.

In addition to the above six broad characteristics of GEC which are implemented in all new development areas of the project, the GEC also follows seven 'key strategic paths' that are taken into consideration as part of strategy planning and the project's key performance indicators. Each strategic path comes with a set of guidelines that are given for a better direction of the project (based on the site visit conducted by Ali Cheshmehzangi in 2016):

- Strategic path 01: Adhering to and maintaining innovation as a development driver in the project;
  Key guideline: including cultivating new impetus to innovation and development.
- Strategic path 02: Adhering to quality-oriented leadership in development;
  key guideline: including exploring new paths for high-end development.
- Strategic path 03: Adhering to green and low-carbon development;
  key guideline: including shaping the new advantages of green development.

- Strategic path 04: Adhering to people livelihood priorities and community-led development;
  key guideline: including building shared development new pattern.
- Strategic path 05: Adhering to keeping an open development (for current and future expansions);
  key guideline: including expanding new cooperation and creating space for win-win opportunities.
- Strategic path 06: Adhering to key tasks of the new development;
  key guideline: including promoting new breakthroughs in harmonious development.
- Strategic path 07: Adhering to continuing to deepen reforms;
  key guideline: including stimulating the comprehensive development of new energy

In order to achieve the project goals, the planning outline was proposed to combine the 'national five major development concept plan' and the 'Shenzhen development strategy plan' to put forward the above strategic paths/approaches for the development of new areas. This outline plan was also addressed in the 'Thirteenth Five-Year Plan of Guangming New District Economic and Social Development'. Moreover, the above key strategic paths demonstrate a clear indication of the project's political status at the national level; an example, with which we can see multiple objectives of eco-development and multi-level governmental actors whom are involved in the whole development process.

Case Study Reflections
While efforts are being made, the city is not yet fully operational, and it is currently on phase two where the majority of the infrastructure and major development projects are in progress. The current project status indicate that the completion date of 2020 is possible and most of the key transportation network are already in place. The GEC's location and its TOD mode are potentially key factors to some of the earlier successes of the project, some of which are based on its rapid GDP growth, industry development, and infrastructure growth. Although described as a striking example in China (i.e., both in research and media), Schwoob et al. (2011, p. 25) argue that 'it is to be feared that the desire for standardisation will prevail over originality'. Similar to some other cases that we have studied in China, real estate projects are now playing a significant role in the GEC project.

Described as one of the ubiquitous eco-city examples by Joss et al. (2013), the GEC project does not represent a unique model of eco-city development in China. Rather, it gives us an overview of a typical national-led model of eco-city. In fact, for the case of GEC, location plays a major role and unlike other similar cases (i.e., in terms of size and project timeframe), such as Cao Feidian Eco-City (Tangshan, in Hefei Province) and Kunming Eco-city (in Yunnan Province), it is accessible to nearby urban and industrial hubs. On the other hand, the TOD mode of GEC allows for more economic vitality and industrial development.

While in the case of Dongtan project in Chongming Island of Shanghai, Arup was working on a complete new development zone, the case of GEC demonstrates more emphasis on a sustainable social/economic community and strikes a balance between modernisation and environmental preservation. The GEC (or as part of the district plan, known as the Guangming New Town Centre) takes an interesting approach by combining urban design and eco-sustainability, arranged into human-scale clusters of housing/farming suburbs in the form of towers/craters conceived as augmentation of existing topography. The project, once completed as planned in 2020, will be a supporting sub-centre to the city of Shenzhen as well as supporting the nearby industries.

5.6 Sino-Singaporean Tianjin Eco-City (SSTEC), Tianjin

Type: International Cooperation

Project Overview

The Sino-Singaporean Tianjin Eco-City (SSTEC) project is a new city development located at the Eastern side of City of Tianjin in China. With its coastal setting, this new development is one of the first eight Chinese eco-cities from the first batch of eco-city projects in China. SSTEC is located 40 km out of Tianjin City Centre and is within the 150 km proximity to the Capital City of Beijing. It is also nearby to Tianjin Economic Technological Development Area (TEDA) and Tianjin Binhai New Area (TBNA), which are two new economic zones of Tianjin. The project is the result of a bilateral agreement between the Chinese and Singaporean governments. The construction of this eco-city project initiated in September 2008 and is expected to be completed by early-to-mid-2020s. When fully developed, SSTEC's expected overall population is estimated to be 350,000 residents within its total area of 34 km². Although nearly half of Chinese urban growth is constructed on arable land, SSTEC is sited on non-arable land, formerly sites for large areas of saltpan, deserted beach and wastewater pond. The overall spatial density in SSTEC is relatively high, at about 10,000 persons per km² (Singapore Government 2016). This is somewhat lower than the projected density in Tianjin's core city, which is anticipated to be 12,500 persons per km² by 2020, but almost double the projection for Binhai's urban core area.

The project started with three key visions to be 'socially harmonious', 'environmentally-friendly' and 'resource-efficient', which all signify the importance of sustainable development. This vision is supported by the concepts of 'Three Harmonies' and 'Three Abilities' (Singapore Government 2016). While three harmonies include 'social harmony', 'economic vibrancy', and 'environmental sustainability', three abilities are focused on the nature of the eco-city being: (1) practicable—including the key elements of affordability and viability of technology use and adaptation in this new development city; (2) replicable—including the knowledge transfer from the principles and models achieved from the development of SSTEC; and (3) scalable—including the application and adaptation of the project's achievements for another development on a different scale (i.e., a larger scale). The project also encompasses two main politically-driven slogans of 'harmonious urbanisation' and 'ecological civilisation' which have both been widely used since 2007, when the SSTEC was also approved for construction.

In July, 2007, The Sino-Singapore Joint Commission set out two key principles for eco-city site selection, which were:

- Capable of demonstrating an eco-city development under the scarce conditions of natural resources, no occupation of arable land and area with limited water resources; and,
- Close to an urban center, capitalizing on the big city transportation and services, so as to minimize the cost for infrastructure development.

Based on the above two selection principles, four possible sites of Baotou (Inner Mongolia), Tangshan (Hebei province), Tianjin municipality and Urumqi (Xinjiang) were singled out for further evaluation. SSTEC's site was finally selected based on the state of existing surrounding infrastructure, the notion of accessibility and connectivity in its region and future commercial capacity and viability. Located in one of the fastest-growing regions in China, SSTEC's close proximity to two main cities of Beijing and Tianjin makes it an ideal location for regional connectivity and economic development. Moreover, there was no arable land within the site. While its economic growth may be overshadowed by the municipal area of Tianjin and two of its viable economic zones, SSTEC can focus on developing a harmonious society which can represent sustainable development in the region.

Project Layout and Planning
There are two unique ecological features of the Tianjin project which differentiate it from the other eco-city projects in China: (1) the eco-city is built in an area lacking water resources and no arable land is occupied; and (2) one-third of this new eco-city is a deserted saltfield, another one-third was polluted land and the rest was wasteland. As urbanisation in China is normally characterised by the expansion into nearby farmland, taking no cropland for the construction of a new city was unprecedented and may provide alternative models for China's urbanisation. The Tianjin initiative also contains other interesting ecological features, such as increasing the use of non-conventional water, expanding the use of renewable energies to power the local economy and dramatically reducing the use of vehicles. It was proposed by the plan that around 25 km^2 of the land should be reclaimed as the new development land in order to accommodate 350,000 residents (Yu 2014). Although lagging a bit behind schedule, more than 70,000 people and over 1000 enterprises have chosen to base themselves in the eco-city by the end of 2016. According to UNEP (2013) it is noted that an upgrade from the currently 31 km^2 area is being considered to a total of 143 km^2, reflecting its high-profile national and international status.

The Sino-Singapore Tianjin Eco-City Administrative Committee (SSTECAC) is the Chinese local authority overseeing the project and exercising political and administrative power. A joint venture, the Sino-Singapore Tianjin Eco-City Investment and Development Co, Ltd. (SSTEC), acts as the master developer for developing infrastructures, residential and commercial real estates. The joint venture is established by a Singaporean consortium led by Keppel Group from Singapore and a Chinese Consortium led by Tianjin TEDA Investment Holding from China (UNEP 2013).

Masterplan

SSTEC's masterplan was jointly developed by parties from both countries. The team includes the China Academy of Urban Planning and Design, the Tianjin Urban Planning and Design Institute and the Singaporean Planning team led by the Urban Redevelopment Authority. Other technical team members were included during the process of project development. This includes key members, such as Bluepath City Consulting team, Siemens and several others, whom have supported the development of KPI system, KPI breakdown and provision of technologies at meso and micro scales. While the SSTEC project is less than half-way to completion, the masterplan endorses new possibilities for small-scale city development in China. With three principle planning elements of 'land-use planning', 'transport planning' and 'green and blue network planning', SSTEC is planned to be compact, mixed use and based on a transit-oriented development (TOD) pattern. Divided in to one central district, two sub-centres and eco-districts, the masterplan of SSTEC proposes for two large central areas of eco-core and eco-chain. There are also six eco-corridors around the new development zone that are expanded towards outside of these districts. All these centres (including the sub-centres) are developed based on a TOD pattern in order to promote a better connectivity between the inner districts and the outer areas of SSTEC. This is proposed under the theme of one axis as a network, three centres and four districts. The axis of SSTEC is referred to as the eco-city valley spine, which goes through all centres and districts and also links the southern and northern surrounding areas of the eco-city.

The most innovative concept in SSTEC's masterplan is the 'Eco-cell' planning idea, which forms the blocks of the eco-city at a smaller scale (Fig. 5.5). Each cell is proposed at a block size of 400 m by 400 m area, including a reasonable walking catchment area to local amenities, transportation and other neighbouring areas. The combination of four neighbouring eco-cells forms an eco-community or eco-neighbourhood. Also the combination of four eco-communities form an eco-district, which enables a better infrastructure development, smart grid pattern and connectivity on a small scale. There are four Eco-districts in the Eco-city.

The use of green corridors and the maximisation of green spaces in SSTEC are key elements of achieving an environmentally-friendly living environments. While all four districts are planned in a compact pattern, the in-between green spaces put more emphasis on developing a harmonious city layout. This reduces the surface coverage for the built environment and proposes for green corridors that link the central green core to the outer green environments.

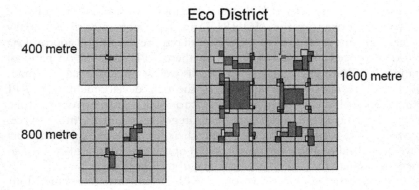

Fig. 5.5 The neighbour layout in SSTEC indicating the concept of 'eco-cell' (adapted and redrawn by the authors, based on the SSTEC documents from the project website)

Key Performance Indicators (KPIs) Guiding Eco-City Development in SSTEC

The concept of eco-city implies that planning for sustainability is to plan our cities in a way to conserve global resources based on ecological principles. At the end of 2007, the Ministry of Housing and Urban-Rural Development (MOHURD) has also issued a policy directive to provide guidance to all local planning authorities on the use of planning key performance indicators in the city-wide land use masterplan preparation and monitoring (MOHURD 2007). In this guidance note, the local planning authorities are requested to adopt a system of key performance indicators (KPIs) for their city-wide Master Plan. This system of indicators has been revised specifically to include new requirements on resource management, such as: water resources balance, water reuse, efficient use of land resources, energy efficiency, emission reduction, as well as recycling activities indicators.

One key point is to translate the KPIs into the masterplanning, and more importantly, into the control plans and then down to individual site/plot level by putting a set of planning constraints. The SSTEC is the first eco-city development model that uses a KPI framework to guide its planning process and attempt to set up an urban management system with support from the decomposed sub-indicators (control targets in SSTEC).

A set of eco-city-related KPIs were developed in April 2008, and were formally approved by the MOHRUD in September 2008. As initially agreed, any future changes to the KPIs will require review and approval via a formal process. The development of these KPIs is based on national standards of both China and Singapore, as well as international best practices that are used for certification of the new development. The KPI framework includes 22 quantitative indicators and 4 qualitative indicators. These are grouped into four assessment categories: environment, resource, economy and society. These KPIs and their decomposed sub-indicators have been used to guide the development of the city's masterplan and also used to evaluate the overall performance of the city.

Based on the SSTEC's division of KPI system into the named four categories, it is required that each category addresses several eco-features. For society, it is indicated that: >90% of trips should be categorised as green trips; there need to be >20% public housing provision; employment housing equilibrium is set at >50%; there need to be coordinated regional policies; and it is also highlighted to include social and culture coordination. For economy, there are two KPIs of having: >50 researchers and engineers per 10,000 labour force; and the regional coordinated economy. For environment, there are several KPIs to achieve: ambient air quality to meet Grade II >310 days per year; Grade IV water bodies; noise pollution to meet 100% of respective functional area standards; zero loss of natural wetland; >70% of preserving the local plant; and towards a coordinated natural ecology for the whole area. Finally, the resource dimension carries the largest load by aiming to achieve the following:

- 100% potable tap water;
- <150 tonnes of carbon emission per US$1 million GDP;
- provision of 100% green buildings;
- provision of >12 m² green space per capita;
- domestic water use of <120 litres per day per capita;
- >60% waste recycling rate;
- free recreational and sports facilities within 500 m of walking distance;
- 100% barrier-free accessibility;
- 100% non-hazardous treatment of wastes;
- 100% coverage of municipal pipelines;
- >20% renewable energy use;
- >50% non-traditional water resource.

The KPI decomposition process includes the breakdown of the 26 KPIs into 51 core factors, 129 key areas and 275 control targets (Fig. 5.6). Here, core factors are referred to the factors that affect the achievability of the original KPIs. For instance, SSTEC requires over 20% renewable energy usage in the city (original KPI), which involves two core factors of 'Increase of Renewable Energy Supply' and 'Reduction of Total Energy Consumption'. Subsequently, key areas refer to the areas that have significant impact on the core factors. For instance, buildings, transport, infrastructure and industry are the four key areas that directly affect the attainability of 'Reduction of Total Energy Consumption'. In addition, control targets are the quantitative and qualitative measurements used to measure performance of the key areas. The control targets are also integrated into corresponding government authorities' routine management. For example, building is a key area to achieve renewable energy usage in SSTEC. Four quantitative control targets are thought to be important for the building sector to contribute positively to the renewable usage target:

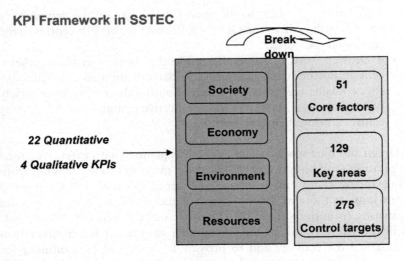

Fig. 5.6 The breakdown of KPIs in SSTEC (drawn by the authors)

- All buildings in the city must be certified by a green building rating system;
- Minimum 70% energy reduction for residential buildings in the design phase and 50% for public buildings;
- Minimum 15–20% operational energy reduction compared to conventional buildings;
- Minimum 5% onsite renewable generation for commercial buildings and 10% for residential buildings.

Green Technologies

SSTEC is developed along a one-axis, three-centres, four-districts concepts. These are described as follows (SSTEC homepage 2016):

- 1 Axis—this refers to the Eco-valley cutting across the Eco-city, which is the green spine of the city. It links up the City Centre, the 2 sub-centres and the 4 districts in the Eco-city, and provides a scenic trail for pedestrians and cyclists. The tram system, which will be built to meet the Eco-city's transport needs, will run along the Eco-valley.
- 3 Centres—this refers to the main City Centre on the promontory on the south bank of the Old Ji Canal and the two sub-centres in the south and the north.
- 4 Districts—this refers to the residential districts in the southern, central, northern and north-eastern parts of the Eco-city. Each district contains several housing neighbourhoods comprising a variety of housing types, as well as their respective commercial and amenity centres serving their communities.

Green and blue spaces—The Eco-city is planned with extensive green (vegetation) and blue (water) networks in mind to provide an endearing living and working environment. The green network will comprise a green lung at the core of the Eco-city and green-relief eco-corridors emanating from the lung to the other parts of the Eco-city. Water bodies in the Eco-city will be linked together for greater water circulation to enhance the ecology and to provide an attractive environment for waterfront development and water-based recreational activities. A wastewater pond will be rehabilitated and transformed into a clean and beautiful lake.

Transport Sector—The transport sector has set a target of 90% green transport. It assumes that only 30% of trips are conducted by walking and cycling. This will be achieved largely through public transport in the form of a Light Rail Transit system, trams and buses. To promote walking and cycling, there will be a 12 km Eco-Valley which connects all major centres and nodes. There will also be community walkways cutting through estates and wide cycling paths on both sides of the roads in the Eco-City (UNEP 2013).

Energy Sector—Energy supply to SSTEC is exogenous from two combined heat and power plants outside the project boundaries. In renewable energy, the main challenge is that there are few sources in Tianjin except for geothermal heat which, together with solar energy for hot water and street lighting, will provide the bulk of renewable energy. Reaching the renewable energy target could therefore become a challenge.

Water Sector—Residents of the Tianjin Eco-City will be able to drink directly from taps. Rainwater will be harvested and, together with recycled water, will be used for landscaping irrigation and general cleaning purposes. With the integration of water-saving technologies into everyday life, residents will also play an important role in conserving water.

Solid waste Sector—To promote effective waste management, residents will be encouraged to sort their waste into categories. Recyclable waste will be sent to recycling stations located within the Eco-City. Non-recyclable waste will be collected via a Pneumatic Waste Collection System and incinerated to generate electricity.

Smart Eco-City Development

The SSTEC was one of the eight national eco-city pilot projects in 2012 and became one of the first batch of national smart city pilot projects in 2013. SSTEC Smart City Action Plan 2013–2015 was approved by the SSTEC Administrative Commission in January 2013. During this period, the new plan was used to guide the development of the smart city in the SSTEC project. The plan proposed a number of projects, which were classified into three categories, i.e., information infrastructure, public information platforms, and applications. It also divided the SSTEC Smart City development into three phases (SSTEC Smart City Action Plan 2013):

- Phase 1 Development Phase (2013): focuses on the architecture design of the SSTEC Smart City and infrastructure construction with a possible budget of 207 million RMB;
- Phase 2 Improvement Phase (2014): focuses on the construction of public information platforms with a possible budget of 168 million RMB;
- Phase 3 Breakthrough Phase (2015): focuses on the area of 'smart ecology', form a smart city application system addressing green government, green industry and green living, with a possible budget of 106 million RMB.

The SSTEC Smart City Plan attempts to improve the city's ecological performance through the adoption of smart technologies. In 2013, the Siemens Corporate Technology launched a research project, led by one of the authors of this book. This research aimed to identify the smart technologies that were being used or planned to use in order to achieve the KPIs in SSTEC, which included: (1) Smart green building technologies—High energy efficiency; building renewable energy use; highly efficient water usage; building energy metering and carbon reporting system; building management system; (2) Smart renewable technologies—Distributed renewable energy generation management; Micro grid energy storage; Power consumption information acquiring system; intelligent-visualised smart grid operation and management system; (3) Smart clean energy technologies—Residual heat reuse; Combined Cooling, Heating and Power (CCHP) and heat pumps; Smart energy management; Comprehensive energy/carbon monitoring and reporting system; (4) Smart green transport technologies—Intelligent transport and facilities management; Traffic simulation tool; Fuel-efficient and low-emission vehicles; the Internet of Things (IoT)- based smart logistic transport; E-mobility; (5) Solid wastes—Solid waste recycling and reuse; IoT-based waste tracking monitoring and management system; and (6) KPIs monitoring system—An IT system focusing on environmental KPIs which can evaluate the performance of KPIs and report annually thus help the work of various governmental agencies

Case Study Reflections

The SSTEC project involves multiple stakeholders as well as a major governmental agreement between the two countries of China and Singapore. This alone indicates a major achievement in linking the local, national and international players of the project. The introduction of the

KPI system and the holistic vision that it offers are used as not only a development instrument but also a monitory system towards achieving a sustainable development model. The current status of the project, being partly completed, cannot fully justify or measure the success of the project. The steady pace of development, however, creates a possibility for researchers and practitioners to assess the effects of such eco-city project in a wider range of practices, mainly in planning and green building design. While the initial expected completion date of 2020 is possibly out of the updated target plan, the eco-city committee's latest update in 2014 indicate further implementation in revising the vision, development of further mechanisms for financial incentives and support as well as enhancement of transport network and the new development's economic viability.

At the masterplan level, several aspects are yet to be addressed. First is the adaptability of SSTEC beyond its first phase and it may sustain in a larger scale, when fully completed or extended. These need to directly respond to current issues of ecological friendly environments, land-use pattern and socio-economic sustainability of SSTEC. Second, the strategy for green connections between the inner and outer areas is yet to play a major role in promoting harmonious planning. This effect can be further promoted through the proposed eco-corridors. Third, the eco-cell planning approach should aim to promote the permeability of the blocks as well as enhancing the connectivity between the blocks. While each cell is conceptually divided in to four smaller sections, this practice should fulfil the detailed design of each of the blocks. And fourth, the project's next phase of development should go under careful revision as it proposes for further development our reclaimed land. This poses risks on ecological quality of the SSTEC project in the future.

The SSTEC project's approach to implementation of both national and international standards in practice puts an emphasis on potential policy and planning transitions at the detailed design level. While on the planning scale, SSTEC offers fewer new plans, some common issues of large block sizes, lack of green corridors and lack of network between the proposed centres are visible as part of the transitions in the overall planning of the eco-city. Here, the common practice of large block layout in China is replaced by the eco-cell planning approach; the introduction of six green corridors, linking outer green spaces and the eco-city's green core, propose opportunities for compact layout for the built environment and the allocation of more space for functional green corridors; and the main network between the four districts and three centres responds to the often lacking network in polycentric city planning approach of China.

Top international developers have launched around 12,000 green homes in the Eco-City for sale with over 5500 homes already completed. The Eco-City has also attracted more than RMB 60 billion in registered capital and around 850 companies, including leading corporations such as Hitachi, Siemens and Phillips, to set up presence in the Eco-City. With a resident population of 70,000 (by the end of 2016) and growing, the Eco-City is springing to life. A range of amenities such as schools, banks, restaurants and shops are progressively opening to service the first waves of residents and companies moving into the Eco-City.

As demonstrated by the SSTEC, win–win public–private partnerships can be harnessed to develop an Eco-City that is practical and affordable (World Bank Report 2009). To achieve this goal, it is important to have good political leadership in partner countries to set the direction and channel the society's resources in the direction of promoting sustainability. A balance also needs to be struck between going for radical, but possibly unaffordable improvements (especially for developing countries) and opting for more cost-effective, practical improvements that can make huge differences on a large scale almost immediately. With the SSTEC, the latter is the approach that Singapore and China have chosen (UNEP 2013)

According to de Jong et al. (2013b), it is noted that the Sino-Singaporean model offered consistent and stable high-level political involvement, substantial public sector investment, and relatively easy cultural understanding, while economic growth figures had also been promising. In this specific case, local governments did not view ecological concerns and economic development as incompatible interests, but rather tried to strike a balance between them.

The key to success in the SSTEC project can be noted as there were multiple joint organisations set up for the project. While the political and administrative power is still held by the Chinese partner (the Administrative Committee), the joint-venture company which is co-managed by both sides is responsible for implementing the project. All of the benefits accruing from operating the eco-city are shared by both parties. More than half of the investment projects in SSTEC come from Singaporean companies, and the Singaporean government contributed resources and expertise to help Singaporean companies locate suitable investment opportunities in the project.

5.7 Sino-Swedish Wuxi Eco-City, Wuxi, Jiangsu

Type: International Cooperation

Project Overview

The Sino-Swedish Wuxi Eco-City (SSWEC) is the result of collaboration between the Chinese local government and several Swedish stakeholders aiming to create a sustainable urban infrastructure comprising environmentally-conscious energy systems (Tan-Mullins et al. 2017, p. 68). This new city-scale development is one of the first batch of eight eco-city projects in China. This project is proposed to create an urban area mainly focusing on the use of new technologies for renewable energies, waste management systems and low-carbon development. According to earlier reports, the SSWEC project is destined to be the home for approximately 20,000 inhabitants. Although the project is still in its initial phase, many of its goals and performance indicators are successfully met. In this project, the integrated systems of sustainability that forms the concept of eco-city, go beyond existing facilities and also comprise of the awareness of sustainability with the people living in it.

The project's ability to succeed is understood to depend on how much its residents embrace the mission that the city has set out to fulfil. One example is provided by the eco-city's transportation. The plan is to assign 80% of the total commuting in the area to public transport, and by doing so, encourage the residents to travel by buses or trains in order to lower total emissions. Approximately, 50% of transportation vehicles are set to run on renewable energy sources (Qian 2011), which is a major target plan in comparison to other projects of the same scale. Another goal for the SSWEC is to exemplify the local inhabitants' cooperation in terms of local waste management. By educating and promoting sound recycling and proper discharge of household organic waste and combustibles, the systems can work in a superior way, where energy, gas and other rest products can be extracted from the waste (Qian 2011; Tan-Mullins et al. 2017, pp. 68–69).

Figure 5.7 shows a conceptual model of environmental strategies adopted in the SSWEC, comprising five sub-models—energy, water, waste, transport and built environment. The aims of this eco-city project, in addition to the promotion of Swedish technologies, is to use the best systems and processes available in order to create an urban environment that minimises its negative environmental impact without affecting the

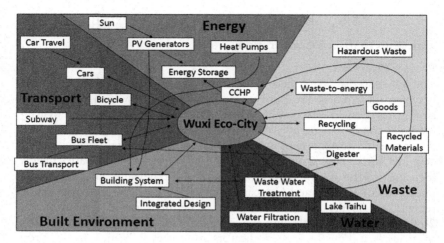

Fig. 5.7 Conceptual model of the Sino-Swedish Wuxi Eco-City (adapted and redrawn by the authors based on the Swedish counterparts 2013)

quality of the city's residents (Tan-Mullins et al. 2017). The ambition is to create an eco-city that serves as both a role-model flagship for eco-cities as well as an educational showcase for future peers (Qian 2011).

Project Layout and Planning

The SSWEC, also known as Wuxi Taihu New Town, is located south of the city of Wuxi in southern Jiangsu Province and Yangtze River Delta.

The overall area is proposed to cover around 150 km², with the buildings covering 95.7 km², and the green lands covering 54.3 km², combining various functions, including government, office, finance and business, culture and entertainment, science and research and, finally, residential areas. As part of the SSWEC, a total area of 2.4 km² is dedicated to the 'national demonstration zone of low carbon eco-city'. The planning is very much based on a Transit-oriented Development (TOD) layout with a central development core area, two major residential districts at the sides and two scientific and R&D parks at the edges of the new eco-city area. The new development is planned on the waterfront with several waster bodies and preserved ecological lands in between (Fig. 5.8).

The SSWEC's target plan is to accommodate around one million residents (and some reports indicate 0.8 million) and 500,000 jobs with a high density CBD at its centre (Wu et al. 2015). The new development

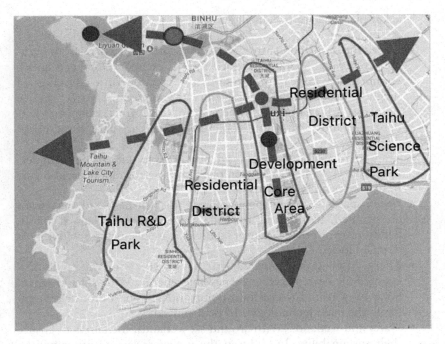

Fig. 5.8 Layout of SSWEC/Wuxi Taihu New Town (adapted and redrawn by the authors based on the Swedish counterparts 2013)

adapted a particular spatial layout, which is divided into three districts: East, Central and West. The masterplanning for the city was led by the Swedish architect firm Tengbom, Export Swedish environmental technologies and management, and used the well-known Swedish eco-city of Hammarby as a precedent. After overpassing the conceptual plan development for the SSWEC, there are two main actions that the government took into consideration: Firstly, they developed a set of innovative ideas to produce a plan which aimed to achieve similar outcomes to those in Hammarby Sjöstad in Stockholm in terms of both ecological performance and economic success. And secondly, the plan was to include a list of target plans as the statutory requirements. Although the seemingly large scale may suggest otherwise, the plan in reality covered an area of 2.4 square kilometres with only amount to 20,000 residents (Wu et al. 2015).

The key planning features include:

- Urban function and urban space—development of a park-front high-end residential neighbourhood, mixed-use redevelopment and a commercial corridor. In addition, the creation of a collection of neighbourhoods each with a distinct characteristic to attract diverse residents. A high-end department store, retail street, exemplary kindergarten and semi-private community green are part of the urban developments;
- Transport—Walkways from residential areas to station and retail destination are planned. A sunken plaza and metro station are also part of the wider mobility plans;
- Green space—A green spine connect adjacent park and waterfront are planned. In addition, roadside green space are planned for. In addition, semi-private interior landscapes are planned for within residential blocks;
- Energy use—It is estimated that about 10.2% of the traditional energy will be replaced by renewable energy. With solar thermal, Solar PV, Ground Source heat pump and Sewage source heat pump being the main technologies to be deployed;
- Water and wetland—With around six wetlands, the aim is to ensure their sustainability. In addition the use of reclaimed water will be emphasised, while the sewage plant will be expanded to handle about 150,000 cubic meter of sewage.
- Green building design—A low-carbon exhibition hall (3 start green building, LEED –NC Platinum) is planned
- Non-conventional water use—There is increased rained water collection and utilisation, planning area is divided into 11 drainage areas with storm water collection facilities. Using infiltration and permeable construction the storm water runoff is guided into a the central green space and the green belt with a storage in a concave green space

Overall, the project developers aims to achieve a comprehensive planning for the project by including various key elements of energy utilisation and solar energy, green building, water recycling, waste treatment; public transport and vehicles with clean energy (de Jong et al. 2013a). The entire recycling system is expected to be completed by the end of 2012. In the meantime, a batch of exemplary projects are in progress, including the development of eco-tech exhibition centre, new energy centre, international school, and eco-community (Chen 2012). Also the project benefits from implementing the certification process. For instance, according to the plan, 70% of buildings should pass the one-star standard of the national green building standard, 20% should pass two-star, and 10% should pass three-star. As per records, until the end of 2013, there were a total of 12 projects (116,600 m² floor space) with Green building design logo. These includes 61,000 m² for three-star, 18,200 m² for two-star and 923,000 m² for one-star. The implementation of such system is continued for any new buildings for the new development.

The SSWEC project also focuses on the mainstream of 'energy' with the following aims (Stoltz 2013): (1) to develop a systems model, taking into consideration all major energy subsystems; (2) to conduct an in-depth analysis of the energy system; and (3) to develop an innovation system model for the eco-city. The conceptual system is to achieve efficient energy systems at five levels, including the building level, the eco-city as a whole new development, Taihu new city in a larger context, the Wuxi region, and the national level. However, this is simply a conceptual framework that is considered to be a part of branding and developing SSWEC as a national eco-city model.

The project also benefits from two tools (proposed by the Swedish team) to assist the eco-city development.

The first tool, called 'Innovation Idea to Deployment Model (EI2D)', consists of 13 qualitative criteria that characterise an eco-project. The focus of these criteria is on a comprehensive idea that includes technical implementation and deployment and how the planning process in an e-city project is to be comprehensively organised, and particularly, from the ecological standpoint. In addition, the EI2D model has been processed and matched against the identified subsystems with the purpose of obtaining additional tools in the matrix of a matrix, where two sets of 'eco-city planning criteria dimensions' and 'eco-city system dimensions (sub-systems)' are introduced. The innovation matrix is used for evaluating the eco-city project, but is also utilised as an aid in the planning of eco-city development. The aim is to keep all the subsystems in mind at each stage of the planning process, in order to avoid so-called 'technological lock-ins' and stimulate innovation as well as synergies. The eco-city planning criteria dimensions are divided into 12 dimensions: (A) Planning and goal setting process; (B) Identification of set goals; (C) System integration; (D) Innovation eco-system; (E) A balanced mix of work, living and shopping/leisure; (F) System management and leadership; (G) Wider energy system access and leadership; (H) Dynamic local and regional energy system; (I) Water supply and water management; (J) Energy efficient buildings; (K) Minimise unnecessary traffic; and (L) Energy Information system. The sub-systems are also categorised into seven dimensions: (1) Energy production and transmission; (2) Energy utilisation systems and the built environment; (3) Urban planning processes; (4) Waste management subsystem; (5) water management sub-system; (6) Transport sub-system; and (7) Energy and information systems (including user behaviour).

The second tool, called 'Energy Target Identification and Deployment Model (ETID)', provides the opportunity to visualise the energy-saving potential of the proposed technology for implementation in an eco-static. ETID has generated scenarios that have been created in connection with the identified sub-systems to reflect different ambition levels for energy saving within the ecosystem. With this, ETID can be a tool for targeting in the planning process. ETID has been developed with the STELLA software and has had the identified sub-systems as a starting point. With an interactive interface, the model aims to assist a wide range of stakeholders in the decision making process. In addition to this, a so-called 'Participatory Approach' was used in the development of the model where a number of key parameters were identified as a result of a workshop with a wide range of invited stakeholders. The actual modelling has been carried out by having inputs from five hierarchical system levels: national, regional, urban, district and building level. The model's evaluations include energy use and CO_2 emissions, which further provide the opportunity to visualize the energy-saving potential of various proposed technologies for implementation in the eco-project. One of the model's main functions is the possibility of active participation of various stakeholders in the goal-setting process in the e-project. This methodology is called 'Participatory Simulation' and aims to incorporate stakeholders' views as quantitative parameters in the model which further serve as validation for the scenarios obtained. Thus, all parameters are resonated and determined and the results visualised. The process is interactive and the result reflects the unified vision of the stakeholders involved (Stoltz and Shafqat 2013).

Case Study Reflections

The construction of SSWEC has started in early 2012 with an expected completion time of 2020. The construction has been slower than initially planned. The Sino-Swedish collaboration can be characterised as a client–provider relationship, but one where the integrated package of solutions is offered with active political support. This approach is valued by the Chinese counterparts of the project, but to a lesser extent than the more comprehensive Singaporean approach (de Jong et al. 2013b), the end-results being that there are no significant stakes of financial investors or academic institutes nor has it led to the establishment of Sino-Swedish joint ventures (ibid.). It is also an exemplar of how it is possible to create values such as housing developments and new employment by adopting a pro-growth planning approach (Chen 2012). Currently, the second phase of the project is under progress and due to completion soon.

A key innovative aspect of the SSWEC is that it is an eco-low-carbon model, which is located as a city within a larger city. Secondly, the planning of the SSWEC was done within the bigger planning of setting up the larger city by introducing a raft of eco-measures into the existing detailed development plan

As the development plan is an official government document, it means the eco-measures were able to be implemented since the project has been initiated. Thus, it provides a new approach of eco-city development through the gradual standardisation of planning requirements. This has the advantage of obtaining central or local government funding and has become one of the key Sino-foreign projects of the first round of eco-cities in China. Furthermore, the project's overall planning was codified by the promulgation of the Wuxi Taihu New town Eco-city Ordinance—the first local legislation for eco-cities in China (Wu 2015). A similar model is seen in some of new joint ventures across the country.

5.8 Conclusions

This chapter has presented six macro-level case studies, categorised in three types of eco-city development, namely 'eco-city with local initiatives', 'national flagship projects', and 'international cooperation projects (also regarded as Sino-foreign projects). Two case study projects represent each of the named categories. Most of these eco-city projects are part of the first batch of macro-level eco-development projects in China. They also represent some of the main experimental eco-city initiatives across the country. Our categorisation is effective in representing the variety of these projects, as we see significant differences in strategies and practices undertaken in individual projects. While the Sino-foreign examples are projects of direct international collaboration, the local initiative and national flagship projects have considered a more localised approach to their development. The main eco-/green features of these projects are mainly embedded in their layout, infrastructure, and spatial arrangements, and how they are implemented in planning practices. Some of these projects have developed a set of key performance indicators (KPIs) while the others have developed a general agenda of eco-development.

At such spatial level of development, as noted earlier (see Sect. 2.3.6), the four dimensions of planning, infrastructure, governance, and policy play significant roles in developing a holistic eco-city project. In comparison to the other two levels of meso and micro (in Chaps. 6 and 7), this requires a more top-down approach from the local governments and allowing for a better utilisation and integration of eco-/green features in the planning practices. Some of these integration models can then be broken down into smaller spatial levels, allowing for a better interplay between spatial levels of the built environment (also discussed further in Chap. 9). More importantly, the macro-level eco-development projects should be able to provide us with a more holistic ecological agenda. In the project examples studied here, although considered in their case study agendas, the role of ecological planning would require to be more effective in the implementation phase. Hence, the approach to macro-level eco-development should eventually shift from these experimental cases of new development and towards more of retrofitting and renewal projects. By scaling up the achievements from these case study examples, we can also propose for new ecologically-friendly masterplans of brownfields and inner-city areas. This, of course, would require a more complex analysis of the existing city environments. However, it is important to recognise the role of these current experimental projects that can later be adopted for future cases of either large-scale projects or smaller scale infill projects in the cities.

Medium to large cities of China have already gone through a phase of redevelopment, and soon some of the early cases would need to revitalise their urban areas. To do so, the eco-development approaches that are demonstrated in this chapter would play a major role in revitalising and restructuring the urban areas. These lessons learnt can set new target plans for cities, and can support the next generation of urban redevelopment in China. A combined method with smart technologies and/or low-carbon agendas will potentially improve the quality of these urban environments. As such, we anticipate to see more hybrid between eco-/green development and other trends, such as smart, low-carbon, resilient, and sponge (Cheshmehzangi 2017). To do so, China needs to re-emphasise on the importance of eco-development on the large scale. Therefore, eco- should not become a sub-matter as we can see already happening in some cases of large-scale projects. This may even require to explore a larger scale of regional planning, through which some of the goals of China's National New-Type Urbanisation Plan (NUP) can also be achieved.

As illustrated in our macro-level cases, not all projects have a set of green technologies. Some of these cases are unique cases (e.g., Chongming Eco-Island) and some are experimental in nature. Therefore, it is important to see how these projects would eventually scale up and their technologies and eco-strategies would then be utilised in other cases at a larger or even a smaller scale. This will then put many of the existing cities into a transitional phase, which we have already seen in the case of Guiyang City. For the past three years, one of the authors of this book has been working on this phenomenon of smart and eco- transitions for the cities and city environments (Smart-Eco-Cities project 2015–18). The more we study various cases, the more we value the importance of transition as a matter of 'necessity' for the cities that are highly polluting or currently lacking in environmental quality. The impacts would then be on development of a circular economy for some, green economy for others, and an eco-development mode for many.

Finally, what we see here is the representation of some of the early eco-city projects of China. These cases represent local initiatives of eco-city models, national flagship projects with the attention of the national government, and Sino-foreign cases of eco-city projects that initiated some of the projects in the first batch of eco-cities in China. Most of these projects are still under development, and hence we avoided to evaluate them in detail. However, it is important to consider these projects based on the varieties they offer. Some of these models are new even at the global level, and some are just at their inception in terms of proposing major planning and land-use reforms in China. To achieve these, we suggest a better integration of multiple dimensions at multispatial levels. The current smart-city initiatives in China may help to better achieve these from the perspective of governance. However, governance alone may not necessarily improve the planning practices. Therefore, what we require is the evaluation of current experimental cases, localising them, and adopting them for the next round of eco-city projects in China. As expressed earlier, some of these may occur at a larger scale, but others can simply be effective at a smaller spatial level of the built environment, namely the district level and also the neighbourhood/community level. Hence, the next chapter will discuss some case studies at the meso level and will further explore the possibilities of eco-development at a smaller scale of community level in China. And later, in the final chapter of the book, we continue to highlight the importance of interplay between these spatial levels. The next few cases, at the meso level, will certainly continue this chapter's discussion of eco-development strategies and implementation.

REFERENCES

Alusi, A., Eccles, R. G., Edmondson, A. C., and Zuzul, T. (2010) *Sustainable Cities: Oxymoron or the Shape of the Future?* Working Paper 11–062, Harvard Business School.

Atkins Global (2013) Future Proofing Cities: Green + Smart, Atkins Lectures, as part of Low Carbon Sustainable Urban Planning Seminar, November 2013.

Birch, E., Abramson, A., Han, A. T., Su, J. Y., and Tongji, C. C. (2014) *Comparison on the Development of Energy Smart Communities among APEC Members*, policy brief as part of APEC Conference on Future Energy Smart Communities Model, October 2014, Penn Institute for Urban Research and Taiwan Institute of Economic Research.

Chang, I. C. and Sheppard, E. (2013) 'China's Eco-Cities as Variegated Urban Sustainability: Dongtan Eco-City and Chongming Eco-Island', Journal of Urban Technology, Vol. 20, No. 1, pp. 57–75.

Chen, X. (2012) *Bilateral Collaborations in Sino-foreign Eco-cities: Lessons for Sino-Dutch Collaboration in Shenzhen International Low-carbon Town*, TU Delft, Delft University of Technology.

Cheshmehzangi, A. (2017) *'Sustainable Development Directions in China: Smart, Eco, or Both?'* in the 2017 International Symposium on Sustainable Smart Eco-city Planning and Development, 11–13 Oct 2017 Cardiff, UK.

Cheshmehzangi, A. and Deng, W. (2015) *Comparison Study of Key Performative Indicator (KPI) Systems in Chinese Eco-City Development*, in the International Conference on Sustainable Urbanisation, Hong Kong, China, 7–9 January 2015.

de Jong, M., Wang, D. and Yu, C. (2013a) *Exploring the relevance of the eco-city concept in China: the case of Shenzhen Sino-Dutch low carbon city.* Journal of Urban Technology, 20(1): pp. 95–113.

de Jong, M., Yu, C., Chen, X., Wang, D. and Weijnen, M. (2013b) Developing robust organizational frameworks for Sino-foreign eco-cities: comparing Sino-Dutch Shenzhen Low Carbon City with other initiatives. Journal of Cleaner Production, 57: p. 209–220.

Franshion Properties (China) Limited (2014) Unleashing Future Vitality of the City, Annual Report 2014. Hong Kong.

Ji, Q., Li, C. and Jones, P. (2017) New green theories of urban development in China, *Sustainable Cities and Society*, 30, pp. 248–253.

Joss, S., Cowley, R. and Tomozeiu, D. (2013) Towards the ubiquitous eco-city: An analysis of the internationalisation of eco-city policy and practice, *Urban Research and Practice*, Vol. 6, No. 1, pp. 54–74.

Joss, S., Tomozeiu, D. and Cowley, R. (2011) *Eco-Cities – A Global Survey 2011*. London: University of Westminster.

Joss, S. (2011) Eco-Cities: The Mainstreaming of Urban Sustainability; Key Characteristics and Driving Factors, International Journal of Sustainable Development and Planning 6:3 (2010) 268–285.

MOHURD (2007) Improving Planning Indicators System, Ministry of Housing and Urban-Rural Development, China Building Industry Press (in Chinese).

Schwoob, M. H., Levan, J., and Rossignol, R. (2011) *China's sustainable cities: players and processes, Annual Report 2011*, by Asia Centre's Energy Program.

SMDRC (Shanghai Municipal Development & Reform Commission) (2010) *Chongming Eco-Island Construction Outline (2010–2020)* [崇明生态岛建设纲要 (2010–2020 年)(摘要)]. Available at: http://www.shdrc.gov.cn/gk/xxgkml/ggjg/zxgg/17344.htm (Accessed on 19-Jul-2017).

SSTEC Smart City Action Plan 2013–2015: Sino-Singapore Tianjin Eco-city Administrative Commision 2013 (in Chinese).

Stoltz, D. B. and Shafqat, O. (2013) Wuxi Sino-Swedish EcoCity – A bilateral Swedish – Chinese EcoCity project with KTH involvement (translated from the Swedish version: Wuxi Sino-Swedish EcoCity – Ett bilateralt svensk-kinesiskt ekostadsprojekt med KTH-inblandning), Published in KYLA + Heat pumps, no. 7.

Tan-Mullins, M., Cheshmehzangi, A., Chien, S. and Xie, L. (2017) *Smart-Eco Cities in China: Trends and City Profiles 2016*. Exeter: University of Exeter (SMART-ECO Project).

UNEP (2013) *The Sino-Singapore Tianjin Eco-City*, in *UNEP South – South cooperation Case Study*, UNEP, Editor. 2013, UNEP: UNEP. p. 10.

United National Environment Programme (UNEP) (2014) Chongming Eco-Island International Evaluation Report, March 2014.

Wang, G. (ed.) (2012) The State of China's Cities 2012/13. Beijing: Foreign Languages Press.

World Bank (2009) *Sino-Singapore Tianjin Eco-city (2009) A case study of an emerging Eco-city*, in *Technical Assistance (TA) report*, No. 59012. World Bank. p. 168.

Wu, F. (2015) *Planning for growth: Urban and regional planning in China*. 2015: RTPI: mediation of space – making of place.

Wu, F., Zhang, F., and Wang, Z. (2015) *Planning China's Future: How planners contribute to growth and development*. RTPI: mediation of space – making of place.

Xie, L., Cheshmehzangi, A., and Tan-Mullins, M. (2017) Planning for Sustainability? A Critical Analysis of Chongming Eco-Island's Planning and Development, presented in the Ecocity World Summit 2017, Melbourne, Australia, 12–14 July 2017.

Yu, L. (2014) Low carbon eco-city: New approach for Chinese urbanisation. *Habitat International*, 44, 102–110.

Zacharias, J. and Tang, Y. (2010) Restructuring and repositioning Shenzhen, China's new mega city, Progress in Planning, 73, pp. 209–249.

5.8.1 Websites

Chongming (2017) available at: www.cmtj.shcm.gov.cn. (Accessed on 19 July2017).

Kohn Pederson Fox (internet source) Meixi Lake Master Plan. Available at: https://www.kpf.com/projects/meixi-lake. (Accessed on 07September 2017).

Gale International (2010) Meixi Lake District. Available at: http://www.galeintl.com/project/meixi-lake-district-changsha-china/. (Accessed on 21 October 2017).

IUCN Environmental Law News (2013) Eco Forum Global Annual Conference Guiyang; Available at: https://cmsdata.iucn.org/downloads/eln2_guiyang_consensus_2013.pdf. (Accessed on 10 July 2017).

Hunan Government (2009) "China Merchants Bank Awards a 3 Billion Credit Line to Help Build Meixi Lake Area," press release, 23 February 2009. Available at: http://www.enghunan.gov.cn/wwwHome/200902/t20090223_152717.htm. (Accessed on 04 October 2017).

McGraw Hill Construction (2010) "Innovation 2010 Big and Super Green: From buildings to cityscapes". Available at: http://construction.com/events/innovation2010/speakers.asp. (Accessed on 21 August 2017).

Qian, Y. (2011) Model city. *China Daily*, 13 May. Available at: http://europe.chinadaily.com.cn/epaper/2011-05/13/content_12505978.htm. (Accessed on 19 July 2017).

Shanghai Municipal Government Webpage (2016) available at: http://www.shanghai.gov.cn/nw2/nw2314/nw2319/nw22396/nw39378/u21aw1101146.htm. (Accessed on 19 July 2017).

Singapore Government on "Sino-Singapore Tianjin Eco-city" (2016) *A model for Sustainable Development*. Available at: http://www.tianjinecocity.gov.sg/bg_masterplan.htm. (Accessed on 19 July 2017).

Smart-Eco-Cities Project (2015–18) Smart eco-cities for a green economy: a comparative study of Europe and China, *Funded by NSFC, project code: 71461137005*, available at: http://www.smart-eco-cities.org/. (Accessed on 06 December 2017).

Stoltz, D. B. (2013) Project description. Department of Energy Technology, KTH. Available at: https://www.kth.se/en/itm/inst/energiteknik/forskning/ett/projekt/wuxiecocity/project-description-1.338870. (Accessed on 21 October 2017).

Woyke, E. (2009) 'Very Smart Cities', Forbes, 3 September 2009. Available at: http://www.forbes.com/2009/09/03/korea-galemeixi-technology-21-century-cities-09-songdo.html. (Accessed on 19 July 2017).

CHAPTER 6

Meso Level: Eco-Neighbourhood/ Community Cases in China

6.1 Introduction

From the perspective of the urban built environment, neighbourhood is a linkage between the individual households and the broad urban context. It is a relatively small geographical unit where social, economic and environmental inter-activities occur. Thus, evaluating a neighbourhood should not only address exactly the immediate neighbourhood environment. It should also include both of the constituent components, such as buildings and people, and a larger spatial context that exerts impacts such as location and urban infrastructure.

Different from the term 'neighbourhood', community is a word, which is people-centric. The term community refers to a group of people that are bound by a specific geography and with a shared sense of collectiveness. Community is transformative and the pivot of information exchange and peer learning. Neighbourhood, as a geographically limited size with intensive interactions between people, is the most important space that people can develop a sense of 'community' and/or 'belonging'.

A 'sustainable community' does not describe just one type of neighbourhood, but the activities that a community engages in to sustain the environment and empower their citizens (Roseland and Spiliotopoulou 2017). One of the prominent examples that are often mentioned for extensive involvement of the general public is the well-known 'Sustainable Seattle Program (SSP)'. Over the past 25 years, SSP has worked to develop

© The Author(s) 2018
W. Deng, A. Cheshmehzangi, *Eco-development in China*,
Palgrave Series in Asia and Pacific Studies,
https://doi.org/10.1007/978-981-10-8345-7_6

five sets of community-based indicators, measuring and making progress towards sustainability in local communities. The development of sustainability targets involved grassroots processes including task teams, civic panel meetings and workshops to form consensus (Sustainable Seattle website n.d.).

Mobilising community is also considered as an effective way in China to move towards sustainability. In 2004, the Ministry of Environmental Protection (MEP) has published the Guidance of Constructing Green Community (GCGC). The Eco-School Scheme was initiated by the Ministry of Environmental Protection (MEP) to encourage environmentally-friendly behaviours in primary schools. In 2016, MOHURD launched a new programme to promote sustainability in the rural area and, at the time of writing, 5855 villages across the country have been certified as 'national green villages'.

There are also neighbourhood-level evaluation systems. Examples include LEED-Neighbourhood Development (LEED-ND), CASSBEE-Urban Development (CASBEE-UD), Green Star—Precincts and BREEAM-Communities. China's specific neighbourhood rating system—Eco-Urban District Evaluation System—was just launched as a new national standards set in October 2017. Earlier on, a neighbourhood-level evaluation system in China—Green Campus Evaluation Standards—has been published in March 2013. These evaluation systems have been used to certify neighbourhood projects in China.

Generally, there are mainly four representative methodologies related to neighbourhood sustainable development:

1. Eco-Industrial Park;
2. Eco-Village;
3. Eco-Business Park;
4. Eco-Residential Compound.

An Eco-Industrial Park is focused on creating a circulation economy by sharing resources and increasing economic efficiency. Currently, around 60 projects have been approved for construction of national pilot eco-industrial parks. An Eco-village aims to create a liveable village and encourage behaviour change by community learning. Its launch marks a recent policy development to address sustainability issues in the rural context. Eco-Business Park is a mixed neighbourhood, often located in city centres or major urban hubs, mainly comprising commercial and office components. Accessibility and walkability are the main issues to be addressed. Eco-residential compounds in China are mostly in the form of small residential districts (SRD), which is a basic planning unit in Chinese cities. SRD is a gated community enclosed by walls and fences. However, there has been a recent debate over opening SRDs to ease urban traffic congestion. There are some business parks and residential compounds certified by neighbourhood evaluation systems, such as LEED-ND and BREEAM-Communities. According to the data from Green Building Map (2017), 22 neighbourhood-sized projects have been certified by LEED-ND in China. In this chapter, we will explore eight case studies drawn from the four neighbourhood categories.

6.2 THE SINO-SINGAPOREAN SUZHOU INDUSTRIAL PARK (SIP), SUZHOU, JIANGSU

Type: Eco-Industrial Park

Project Overview
The Sino-Singaporean Suzhou Industrial Park (SIP) is a strategic economic partnership between the Chinese and Singaporean governments. Launched in 1994, it was the flagship cooperation between both parties for the development of an innovative eco-industrial model park and is situated in Suzhou City of Jiangsu province China (Sim 2015). The initial conceptualisation of the project is attributed to the former Chinese leader, Deng Xiaoping, for his comments on the need for China to leverage on experiences from Singapore in key areas of economic and social development, during a tour of southern China in February of 1992. Seeing China's interest as a potentially beneficial opportunity to both parties, Lee Kuan Yew, the then Singaporean senior minister, called for a bilateral project during a state visit to China in September/October 1992. This led to the signing of the SIP agreement between the Singapore Labour Foundation (SLF) International, and the Suzhou government in December 1992 (Souza 2009; Pereira 2003).

The Suzhou Industrial Park (SIP) employs a three-pronged collaboration approach. At the topmost level is a Sino-Singaporean Joint Steering Council (JSC), which provides strategic direction and leadership for the project. The next level is a joint working committee whose members include the mayor of Suzhou city, chairman of the Singaporean Ministry of Trade and Industry, key government officials from Jiangsu province,

Suzhou city and Singapore, who meet around four times a year to deal with more operational issues in the SIP. The bottom level involves more operational daily leadership provided by the joint venture company, the Sino-Singaporean Suzhou Industrial Park Development Corporation Ltd. (CSSD), and the Suzhou Industrial Park Administrative Committee (SIPAC). The ownership of the SIP was based on 65% for Singapore, while the Chinese parties held 35%, until these figures were reversed in 2001. As at 2012 the Singaporean parties holds 28% within the Sino-Singaporean SIP development corporation Ltd (CSSD), while the Chinese parties hold the remaining stake (de Jong et al. 2013).

The SIP is located in the beautiful city of Suzhou, in the Jiangsu province of China. Suzhou is a key traffic hub with relatively good connections from the Shanghai–Nanjing Expressway, the Suzhou–Jiaxing–Hangzhou Expressway, the Suzhou–Shanghai Expressway, the Shanghai–Nanjing Railway, the Shanghai–Nanjing Intercity Railway, and the Beijing–Shanghai High Speed Railway. Intended to be the new eastern township of Suzhou city, the SIP covers a total administrative area of 288 km^2, of which 80 km^2 is devoted to the China–Singapore cooperative zone. As an industrial park, the SIP boosts of a number of layouts, including: the industrial level (machine manufacture), electronics, information, modern service industry, nano-tech applications, bio-medical sciences, and cloud computing. The SIP consists of three key districts: Jin Ji Lake rim CBD, the Du Shu Lake Innovation District of Science and Technology, and the Yang Cheng Lake Tourism Resort. It also has two gateways: the SIP Integrated Bonded Zone and Donghuan Road Area. Furthermore, it is connected with the Suzhou Dushu Lake Science and Education Innovation District (SEID), a 25 square kilometre planned area for universities, science and technology platforms (SIP website).

Eco-Initiatives and Green Features

The SIP is one of the few industrial parks to have actively sought to blend traditional culture with modern advances, particularly with regards to urban environmental issues. It currently has more than 25 million square metres of green area, and records 45% green coverage rate. The annual output of its over 100 energy-saving and environment-protecting industrial projects totals around 42 billion RMB. Key eco-initiatives within the SIP include (SIP website):

Energy and Waste Management: There are 76 two-star International Green Building projects, and a dozen Gold LEED-certified buildings in the SIP. Its recycling system consisting of wastewater treatment plants, sludge drying facilities, thermal plants, and central AC system covers the sewage network across the entire area. The SIP Sino French Environment Technology Co. Ltd, a joint venture which 49% is owned by Sino French Water and 51% owned by China Singapore Public Utilities, operates the first sludge treatment plant designed in Jiangsu province designed and built by Degrémont.

Transportation: Pedestrian walkways, bicycle lanes and functional public bike services are prevalent across the SIP. Also its smart bus system is capable of scheduling and managing 640 buses, as well as releasing real-time bus information at 455 bus stops.

Social-Cultural: A number of social-cultural activities are often held within the SIP. Communities and schools hold flea market and car-free days, publish books, teaching materials and pamphlets promoting a low-carbon lifestyle. 58 residential communities try out garbage classification; 25 schools pair with local companies in protecting environment; the green coverage rate in residential communities is 90%, and 95% schools are recognised as green campus; and the former rural townships of Loufeng, Weiting and Shengpu are certified as beautiful towns by the national authority. Furthermore, the SIP has a range of facilities, including housing, retirement villages, recreational and senior care facilities, as well as schools, including the Suzhou Singapore International School.

Case Study Reflections

Despite having achieved much in terms of development, SIP faces several key challenges. For example, low-end and labour-intensive industries such as electrical manufacturing are being replaced by high-tech and service industries like nanotechnology and finance. But analysts say there should be an eco-system of large firms and small vendors. Otherwise, transport costs could rise if small firms cannot enter or have to move out. There is also a need to further streamline the SIP's bureaucracy and to better examine officials for top jobs after two officials were placed under anti-corruption investigations.

In two decades, the authorities have turned down 300 projects with total investment of US$ 2 billion since they posed potential hazards to the local environment. SIP leads among national development zones with 351 ISO14001-certified companies. As a result, Its COD (Chemical Oxygen Demand), carbon dioxide emission for producing 10,000-yuan GDP and comprehensive energy consumption in 2013 is one-eighteenth, one-fortieth, and one-third, respectively, of national averages. By October 2015, a total of 5512 foreign businesses were reported to have been attracted to the SIP amounting to a contractual investment of EUR 37.58 billion while import and export volume was totalled to EUR 5658.73 billion (EU SME Centre website). The SIP is listed as one of the country's first model ecological industrial parks, and pilot areas in building ecological civilisation. The SIP project represents one of the early eco-industrial parks of China. It also highlights the importance of industries and low-carbon transitions in the process of development.

6.3 The Sino-German Qingdao Eco-Industrial Park, Qingdao, Shandong

Type: Eco-Industrial Park

Project Overview

The Sino-German Qingdao Eco-Industrial Park (SGQEP) is a strategic economic cooperation between the Chinese and German governments, focused on the areas of ecological and sustainable development. The project seeks to leverage on the technical know-how of both parties in the application and demonstration of environmental standards and technologies in fields such as energy technology, automotive, automation and supply industries, and in the bio-pharmaceutical industries. In particular, it aims to enhance industrial and technological development by attracting both German and other foreign technical and service-based firms to the eco-park. Development of the SGQEP was approved by the state council of the government of the People's Republic of China on 4 January 2011, as part of the 'Blue economic zone' development plan in Shandong Peninsula (Sino-German EcoPark website; Xinhua 2017).

The SGQEP is situated in Huangdao district, 30 km from Qingdao Lui'ting international airport and 6 km from Qingdao bay ports in the port city of Qingdao, Shandong province, China. It sits on a land area of 11.6 km^2, towards the northern fringe of the economic and technological development area in Qingdao. The project design and implementation involved collaboration amongst key parties, such as the Chinese Ministry of Commerce as the key stakeholder, the German Federal Ministry for Economy and Technology, the City of Qingdao (which has a substantial German cultural heritage, being a former German protectorate between 1897–1914), and technical partners from China and Germany providing support in the areas of masterplanning, design and sharing of experiences in technologies and eco-city standards. As an industrial park, the SGQEP boosts of a number of layouts, including: ecological residential areas;

industrial areas covering both German and Chinese manufacturing companies; science and technology research facilities and schools; with good access to other amenities within the main city of Qingdao (Sino-German EcoPark website).

Eco-Initiatives and Green Features
Central to the conceptualisation of the SGQEP was the idea of 'green integration' in key areas involving planning for the urban spaces, building design, traffic management and landscaping. Of its 11.6 km² land area, 30% was allocated for green spaces and landscaping, while the industrial and residential areas used up 45% and 25%, respectively. Below we briefly discuss key eco-initiatives within the SGQEP (Sino-German EcoPark website; Siemens 2017; Yang et al. 2017; Ling et al. 2013; German Energy Center & College website n.d.):

A Passive House Demonstration Project: Occupying a land area of 13,768 m² sits a passive house technology centre within the SGQEP, with facilities ranging from showrooms, multifunctional meeting rooms and office spaces to residential apartments. The project, which was developed by Siemens, serves as a demonstration of the importance and principles of passive house technology in building energy management. Passive house technology is primarily founded on a set of building standards, which stipulates key requirements in terms of architectural planning, technology usage and ecological planning. It leverages on thermal insulation and the use of heat exchangers in moderating energy usage within buildings. Available figures show that the passive house technology demonstration centre saves about 1.3 million kWh of energy annually, translating to 500,000 RMB in energy cost savings, and that it uses 90% less energy than a building of similar size and capacity.

Energy Efficiency in Building Development: In addition to the passive house technology centre, other industrial and residential developments within the SGQEP were built with a strict adherence to energy efficiency standards. For example, the residential areas within the SGQEP, accommodating 7685 inhabitants, were developed in line with China's two-star system for green buildings. This standard, comparable to the LEED ratings for buildings, evaluates building development with respect to factors such as: water and energy savings, materials savings, indoor environment quality, land savings and outdoor environment, amongst other factors. Industrial developments within the SGQEP also adhered to strict energy efficiency standards, with the German enterprise centre recognised as the first DGNB gold-certified office building in China.

Ecological Primary School Demonstration Project: Eco-initiatives were also extended to the creation of a model eco-primary school within the SGQEP. The project covers a land area of 3.5 hectares, with the current student population put at 1620. Developed to achieve a zero-energy-consumption standard, it is reported to be the first eco-primary school in China and seeks to serve as a model for the adoption of ecological technology standards within school developments in China.

Case Study Reflections
Since its initial conceptualisation and eventual development, the SGQEP has continued to witness tremendous improvement in investments and interest from both foreign and local firms. Figures, as at the first quarter of 2016, placed investment volume within the SGQEP to about US$51 million. Notwithstanding the reported successes discussed above, concerns still remain with respect to the ecological and sustainable development drive of the SGQEP. For example, support from both the local and national government in China helped open up the SGQEP to Qingdao main city, with the construction of roads and linking bridges due to its remote location. As a consequence, there was an increase in the number of cars, which were initially controlled within the SGQEP, leading to concerns over the level of vehicle emissions (Ghiglione and Larbi 2015). This factor reflects on the planning system that is implemented in the project.

Questions of the economic viability of the SGQEP, vis-à-vis its remote location from Qingdao downtown, is seen as the main reason for the possible trade-off in allowing the influx of cars within the SGQEP. With plans in the works for a subway line and other public transport measures to connect the SGQEP with Qingdao main city, it is hoped such issues will be addressed. However, this only depends on how this new area is accessible in the future for trade, businesses, and new industries. It is anticipated that it would still remain dependent on the industries in the surrounding areas.

6.4 Tengtou Village, Ningbo, Zhejiang

Type: Eco-Village Development

Project Overview
Tengtou is a small village settlement located in Fenghua district within the City of Ningbo, in Zhejiang province: an eastern coastal province of China. It sits on a land area of two square kilometres, with a current

population of 817 people. Famous for its eco-tourism attraction potential, Tengtou residents are mostly engaged in agriculture and tourism-focused occupations. Fenghua district itself is a culturally and politically important part of Chinese history. Established as a county during the Tang dynasty, it is the ancestral birthplace of Chiang Kai-shek, who served as the leader of the Republic of China between 1928 and 1975 (Fenghua Government website 2017; Top China Travel website 2017). The region, therefore, offer a combination of tourism, political status, and historical values.

Tengtou village is famous for its eco-friendly developmental approach. This is widely acknowledged in the literature, making Tengtou the only village selected to be exhibited in the Urban best practice area of the Shanghai Expo in 2010. It has also been recognised by the United Nations as being amongst the 'Global Ecological Habitats 500', and is listed as No. 6 in China's Top Influential villages (Han 2010; China Net 2010).

Eco-Initiatives and Green Features
From the onset, developmental initiatives within Tengtou village were planned with keen consideration and adherence to environmental preservation. Particularly during China's reform and opening-up period during the 1980s, Tengtou village had set out clear pathways for its development, some of which includes in the followings (Han 2010; China Net 2010):

Environment-Friendly Development Assessment: Developmental initiatives within Tengtou village are put through a rigorous assessment process, in order to determine their environmental impacts prior to project approval and commencement. For example, a proposal to build a pulp and paper mill in the village by a foreign investor in the 1980s, promising an impressive annual profit of over 1 million RMB, was rejected by the local authorities after its impact on the environment was further assessed. To get the buy-ins of the villagers, a documentary was screened, with the help of an environmental specialist, detailing the impacts of pollution and environmental mismanagement on the occurrence of future environmental and health disasters. As a consequence, the local authorities and the villagers reached a unanimous decision to cancel the project. In furtherance to project assessment in Tengtou village, an environmental protection committee was set up in 1991. Being the first such village-level environmental protection body in China, it was aimed at assessing and approving projects for development within the village. It is reported that over 50 projects, with impressive economic returns but high environmental impacts, have been rejected in Tengtou village.

Renewable Energy Adoption: Tengtou village broadly adopts renewable technologies for environmental protection, with ecological conservation funding by local authorities growing by about 20% annually. The village saves about 50,000 kWh of electricity annually from its residential communities, with these communities equipped with solar-powered water heating systems. An estimated one-third of its roads are reported to be equipped with both wind- and solar-powered street lighting. In addition, the data on air quality, temperature, and humidity conditions within the village are constantly monitored and displayed on electronic boards. These data are measured by the local air-quality monitoring station, helping the villagers to stay abreast of real-time conditions. Therefore, there appears to be a strong sense of awareness of climatic conditions and air quality in the community.

Water and Waste Recycling: Tengtou village broadly practices waste recycling and also encourages the use of fabric bags, as opposed to plastic bags, for shopping. Rainwater and sewage wastes are separately managed within residential areas: Tengtou's main waste and garbage system is managed using an eco-friendly process, within the main Fenghua district. The village saves an estimated 9500 tonnes of water annually by using captured rainwater within its two eco-friendly toilets constructed for the village.

Agricultural Innovation and Collaboration: In addition to eco-tourism, agriculture serves as a key occupation for the residents in Tengtou village. Prior to the intervention of the local authorities, the villagers had problems engaging in farming activities due to the unfavourable natural conditions of low altitudes and farmlands being submerged from rainfalls. To address the issues the local authority bought up the farmlands, founded an agricultural company focused on developing an industrial agriculture programme, and adopted three key practices: Firstly, they employed organic farming practices whereby the use of pesticides and other artificial fertilizers were discouraged; secondly, a three-dimensional eco-farming approach was adopted wherein farming was practiced under a three-layered framework. The upper layer involved the cultivation of grapes, with this layer serving as a shield against heat and sunlight. The middle layer was used for breeding birds, with the fallen grapes from the first layer serving as a source of food for the birds. While the last layer was used for fish farming, with the fishes feeding on fallen bird wastes from the middle layer. This ecological approach to agriculture was highly commended by officials from the United Nations, earning Tengtou village the Global 500

Award for Environmental Achievement; and lastly, the local authorities in Tengtou village embarked on a number of project collaborations and cooperation with research institutions and universities for the advancement and adoption of technologies for agricultural efficiency. As a consequence, plant tissues and cultivated seedlings from various plantation-yielding fruits were developed.

With such focused initiatives, Tengtou village has seen a huge increase in its agricultural outputs and earnings. It is reported to have earned 30 million RMB from supplying flowers and plants for the Beijing Olympics event in 2008, and also provided the 2010 Shanghai World Expo with about 100,000 tree seedlings. Also its agricultural success and expansion as a major industry has seen an increased demand for land resources, with villagers relocating to other cities and/or provinces within China to set up plantations.

Case Study Reflections
Tengtou village continues to draw increasing attention for its eco-friendly development from tourists and public officials in different parts of China and the world. Its plantations, particularly with beautiful gardens and flower fields, are key tourist attractions. Tengtou is known to be the first village in China to sell tickets to visitors since 1999. Figures in 2009 place ticket sales around about 26.3 million RMB, and the overall tourism earnings were estimated to be about 119 million RMB annually. Furthermore, Tengtou village's model is being studied for its eventual replication to other parts of China, and possibly the world: the village has seen an influx of government officials from various parts of China to study and share experience with Tengtou local authorities on its success. In addition, a mimicked version of Tengtou village, called Tengtou pavilion, was built and displayed at the 2010 Shanghai Expo, with visitors at the expo treated to a typical rural experience of China.

There is still much to be accomplished in Tengtou village with respect to its drive for ecological conservation, enhancing the quality of life of the villagers and its management. Nevertheless, Tengtou's proactive approach to eco-development and its promotion of eco-tourism as a key economic driver provides a valuable lesson for other cities/towns aiming for ecological conservation. Tengtou sets as a remarkable example of its kind amongst Chinese villages that is actively focused on eco-tourism and ecological conservation measures.

6.5 Gubeikou Township, Beijing

Type: Eco-Village Development

Project Description

Gubeikou Township is a strategic water town, supplying most of Beijing's drinking water sources. It is located within Miyun County towards the northeastern part of Beijing Municipality. To the north, it is linked by Luanping County of Hebei province, and to the west by Gaoling and Xinchengzi (both of them towns within Beijing municipality). Gubeikou previously served as an important military passage, connecting much of the northern and southern parts of the Yanshan range. It covers a total area of 85.82 km^2, with jurisdiction over nine villages and four residential neighborhoods. As at 2016, available data placed its total population to about 9154 people, of which 6834 are primarily farmers, accounting for 75% of the total population. This township is not necessarily a neighbourhood-scale project, but is one of the case of eco-village development in China.

Gubeikou is planned to be functionally liveable, environmentally friendly, and economically prosperous, with its economy driven by a high degree of tourism-oriented culture. In 2008, the Ministry of Finance (MoF), the Ministry of Housing and Urban–Rural Development (MOHURD), and the National Development and Reform Commission (NDRC) selected the town as one of the first seven national pilot towns for green and low-carbon development. At the same time, it was also recognised as a key Chinese historical and cultural town (Beijing Miyun 2017; Beijing Tourism 2016; Gubeikou Government website n.d.).

Eco-Initiatives and Green Features

Planning initiatives embarked upon by Gubeikou Township is geared towards achieving three key targets: (1) a green and environmentally friendly town; (2) a culturally preserved zone; and (3) a livable community. Moreover, from the planning perspective, Gubeikou Township is divided into three ecological areas: the first includes mountainous areas, focusing on the conservation of mountainous ecological systems combined with the moderate development of eco-tourism; the second area includes the town centre, focusing on ecological town construction and energy-efficient buildings; while the third is the Chao River, which is a strip of green space with a width of between 100 and 200 meters designed to form a transition between the mountainous area and the town center, and also to serve as a public open space to create interactions amongst people.

In 2012, Gubeikou further developed a plan specially for constructing green and low-carbon township, which proposed a set of targets that must be met as at 2015. Among the key environmental targets are: a 25% carbon reduction per GDP unit between 2010 and 2015; ensuring that all public and new buildings meet the national building energy efficiency standard or attain green building certification; that 30% of households should use at least one renewable technology as energy source, and to attain a 95% mark for wastewater treatment. We briefly discuss progress made towards these targets (Gubeikou Government website n.d.; Gubeikou Township Government 2008; Beijing Miyun 2017):

Promote the use of renewable technologies: From 2010 to 2015, during the period of the 12th Five-Year Plan, Gubeikou has gradually increased the use of solar energy, shallow ground energy, and biomass as the source of energy for buildings. The main technologies adopted are BIPV, ground source heat pump, water source heat pump and reuse of waste heat. All roads and streets have been installed with solar street lamps, and 64.5% of households have been installed with solar hot water systems. Four biomass-fuelled gas stations were built to provide gas to 30% of the households. The street lamps along the main road have also been replaced with solar and/or LED lighting.

Promote green building development: All new buildings in Gubeikou were encouraged to achieve the national mandatory energy efficiency standard for the cold climates. Through this promotion process, developers were also encouraged to attain a GBES certification for their projects. The compliance rate to the national building energy standard is still low in rural areas. Gubeikou is one of the few townships in the rural area that put energy efficiency as a compulsory requirement. The largest new residential development in Gubeikou is Shimatai Village. In addition, air source heat pumps were installed to provide heating in wintertime, saving about 3000 tons of coal compared to heating sourced from coal burning.

A large percentage of residential houses were originally built with no insulation, meaning that they were very susceptible to heat loss. In addition, they were very expensive to heat. The town has invested 18 million RMB to insulate those houses since 2012, and by 2015, 820 houses, totalling 64,000 m² of floor area, has been upgraded with better insulation and other energy-efficient measures, such as installing thermal-protective doors and windows. Government buildings and school buildings (in the category of public buildings) were also retrofitted with better insulation, and efficient heating provisions.

Improved environmental quality and community facilities: Gubeikou, as previously stated, is a strategic water town. A key issue, however, has been the contamination of its waterways from the disposal of sewage. To address this issue, a central wastewater treatment plant was built in the township and 23 small-sized treatment stations in the villages. A waste transfer station was also built, as a temporary stack site, before wastes are moved to the disposal center in the Miyun County. Furthermore, investments were also made for the provision of community-based facilities. About 20 million RMB has been invested in building a cultural and sport centre, a kindergarten and a service centre. Every individual village has also been equipped with a square and sports facilities installed to foster and enhance interaction amongst residents.

Case Study Reflections
Gubeikou Township continues to enjoy tremendous improvements in terms of its eco-development agenda. It is also attracted attention from both local and national governmental levels, given its strategic importance in serving as Beijing's drinking water source. Various projects scattered across its township area focused on creating environmental, cultural, and liveable environments, which are already attracting an increasing number of tourists. For instance, in 2015, more than 1.47 million tourists were reported to have visited the nine villages and four residential neighborhoods under its jurisdiction, leading to an income of 460 million RMB; while figures from 2016 place the number of tourists visiting at 243 million people drawing an income value of 720 million RMB (Gubeikou Government website n.d.). Gubeikou is currently listed as one of the most beautiful leisure townships within the Beijing municipality, given its rapid adherence to eco-environmental initiatives while at the same time maintaining its traditional and cultural foundations.

The project is still under further development to achieve its eco-environmental initiatives and agenda. Alongside the recent innovation-driven development region in the Xiong'an New Area (announced in autumn 2017), Gubeikou still remains one of the important areas in the region. However, the project's focus, particularly in the past few years, has been mostly on tourism industries, which puts some pressure on their initial environmental protection strategies. As a result of its natural importance in the region, it is hoped this township area further focus on balancing between the nourishing tourism industries and the environmental protection initiatives.

6.6 The Sino-Singaporean Guangzhou Knowledge City: Ascendas OneHub Business Park, Guangzhou, Guangdong

Type: Eco-Business Park

Project Overview

The development of the Sino-Singaporean Guangzhou Knowledge City (Herein referred to as SSGKC) represents the third strategic urban city development project between stakeholders from both China and Singapore. The project is focused on the areas of economic development, knowledge exchange and urban sustainability. Unlike prior projects (i.e., the Sino-Singaporean Suzhou Eco-industrial Park (SIP) and the Sino-Singaporean Tianjin Eco-city), which were driven by government-to-government collaboration, the SSGKC is led by private sector collaboration with substantial government support from both countries. The SSGKC, located in Guangzhou city within China's prosperous Guangdong province, spans a total land area of 123 km^2 and was developed to serve as a centre for knowledge-based industries focused on areas such as communications technology, clean energy, pharmaceutical and biotechnologies. It was primarily aimed at repositioning Guangzhou's economy, which had been a manufacturing and labour-intensive economy, to a knowledge-driven economy by attracting and warehousing knowledge-based industries and global talents within the project (Guangzhou Knowledge City Project 2016; Ti 2015; Singbridge 2016).

Embedded within the SSGKC is the 'Ascendas OneHub' eco-business park, jointly developed by Ascendas-Singbridge Group from Singapore and the Guangzhou Knowledge City joint venture company. The project sits on a land area of 30 hectares and involves a one-stop mixed-use development consisting of new spaces for commercial, residential, and recreational developments. Leveraging on the idea of 'work, live, play', the Ascendas OneHub Eco-business Park seeks to integrate research-led organisations, multinational corporations, and global talents within a knowledge-driven and innovative community (Ascendas OneHub GKC 2016).

Eco-Initiatives and Green Features

The Ascendas OneHub eco-business park project was designed by S333 Architecture+Urbanism Ltd, a UK-based architectural firm, employing international best practices in green building development and urban sustainability. The new development is done in line with the US-based green standards (LEED) for building design. Key eco-initiatives embarked upon within the park include (Singbridge 2016; Ascendas OneHub GKC 2016; Sino-Singapore Guangzhou Knowledge City 2017):

Green Space Development: Adopting Singapore's famous 'Garden City' model, the OneHub Eco-business Park is strategically located on a beautiful green valley within the SSGKC. It also reflects the beautiful surrounding natural environment. It is designed with breathtaking landscapes, and an estimated 45% of land use within the park are devoted to green spaces.

Transport Network: The business park offers appropriate accessibility to connecting traffic routes. It sits on the main traffic road within the SSGKC and is positioned close to the Guangzhou Knowledge City metro line currently under construction, which further connects this new business park to Metro Lines 14 and 21 of Guangzhou City. A public bus interchange is also under construction, which it is proposed should be directly connected to the offices within the business park. This potentially helps to promote the use of public transport systems, thereby reducing the carbon footprint from reliance on privately operated cars within the park.

Research Development Initiative: The project also promotes research on technology development and focuses on key areas of urban sustainability and development, pollution control, biomaterials, electro mobility, and food sciences. To support this, a Sino-Singaporean International Joint Research Institute (SSIJRI) was established in 2015. The SSIJRI, a collaboration between Nanyang Technological University of Singapore and the Southern China University of Technology, currently runs about 23 projects with funding support of about 250 million RMB from the Guangzhou City government and the Guangzhou Development District. The SSIJRI's focus areas target mostly projects in related to urban sustainability issues. This initiative provides a collaborative platform for ensuring eco-development within the park, and by extension to the SSGKC development zone.

Case Study Reflections
The Ascendas OneHub Eco-business Park holds a strategic position in the SSGKC's ambition to serve as Guangzhou's pioneer for a knowledge-driven economy. It presents a major meeting point for businesses hoping to tap into, and contribute to, the potential presented by such economic

shift. Though a lot of the projects within the park are still under construction, initial commitments shown so far with respect to investments, regulatory frameworks, government support, and interests from multinational corporations, show that progress is being made. However, what remains to be seen is how sustainable the project will be in the face of changing and conflicting requirements between achieving urban sustainability and economic viability as was the case with the Sino-German Qingdao Eco-Industrial Park (SGQEP), where a trade-off was made to loosen the control on the influx of cars into the SGQEP so as to open up the SGQEP to Qingdao main city. Moreover, the project's location and its relationship to the municipality of Guangzhou makes it important as a new zone in this economically viable region. It plays a major role in development of innovative industries as well providing support to surrounding industries in the region.

6.7 Hongqiao Business Park, Shanghai

Type: Eco-Business Park

Project Overview

The Hongqiao Business Park or central business district (Herein referred to as HQCBD) was established in 2009 as a strategic initiative by the Shanghai municipal government to enhance and promote the city's business environment. The project integrates four key districts: Changning, Jiading, Minhang, and Qingpu. It also provides an alternative business development environment within the municipality of Shanghai, away from the crowded main Pudong Central Business District (CBD). Its proximity to the Hongqiao transportation hub, consisting of both an airport (with domestic and international terminals) and a high-speed railway station, makes it even more important as a major business hub in the region. HQCBD is located towards the western fringe of the Shanghai Pu Dong CBD, and covers a land area of 86 km^2. Of its total land area, 59 km^2 is reserved for future expansion initiatives, while 27 km^2 was earmarked for immediate development with 4.7 km^2 dedicated to the development of the core business area consisting of commercial office spaces, limited residential apartments, hotels, retail and shopping centres, recreational, and cinema facilities (Thomas 2012; Shanghai Hongqiao Central Business District 2016).

Eco-Initiatives and Green Features

Local authorities in Shanghai put considerable emphasis on the urban sustainability agenda in the planning and development of the core business area within the HQCBD. A mix of high-performance building insulation, natural ventilation, green roof technology, and water recycling practices, amongst others, were adopted for the development of the HQCBD. In the highlights below, we briefly discuss two of the main eco-initiatives embarked upon within the park (Ramanujam 2015; de Vries and Dai 2015; Shanghai Hongqiao Central Business District 2016; Amy 2013; Tang 2016):

Green Building Development: A strategic eco-partnership was agreed upon between the local authorities and the US Green Building Council (USGBC) for the creation of a low-carbon community within the HQCBD. This involved the adoption of LEED certification for green buildings, and also ensures compliance to China's green building standards within the HQCBD. As a consequence of this major initiative, a large number of building projects within the HQCBD have applied for LEED certification, while new targets have been set to ensure every project within the HQCBD satisfies the requirement for China's two-star system for green buildings. This approach sets a remarkable example in the region for large scale green building certification.

Low-carbon Transportation Network: The development of the HQCBD was integrated with Hongqiao's extensive transportation network, which consists of the Hongqiao International Airport, metro lines, high-speed rail corridor, inter-city and intra-city bus lines. The transport network serves as a hub for local, regional and international connections for residents and businesses within the HQCBD, with the hub's passenger flow reaching 250 million in 2013. This integration enhances the ease of travel and encourages the use of sustainable modes of travel, thereby minimising travel-related carbon footprints. The methods undertaken are to enhance the multilayered transportation network of the area, by focusing on the possibilities for low-carbon transportation. The emphasis is given on the integration of the already existing transportation network in the new development, and also to enhance it where needed. This extension allows for improved accessibility to public transportation means and other connections in the larger context of the Shanghai Municipality. The project also sits in close proximity to the industrial regions around Shanghai, allowing for more transit-oriented development opportunities in the larger context. Therefore, the emphasis on low-carbon transportation network is expected to be a major mainstream of the HQCBD development.

Case Study Reflections

With the development of the HQCBD, it is hoped that businesses operating within Shanghai's main Pu Dong CBD will consider relocating in order to take advantage of lower rental fees and business operating costs. However, as the project progressed, concerns were raised with respect to the cost of relocation regarding the potential benefits promised, and calls made for inducements in the form of tax rebate for companies considering possible relocation (Thomas 2012). Reports show that the investment agreement between local authorities and about 28 companies to establish offices within the HQCBD totalled 11.7 billion RMB as at 2016, with about 220 local and international companies currently operating within HQCBD. From a sustainability perspective, the HQCBD's urban transport integration initiative and its wider sustainability goals embody a positive model to emulate for future business park developments.

The HQCBD project also incorporates some of Shanghai's smart city initiatives. For instance, to further develop the low-carbon transportation network, the project is now preparing for the creation of a new comprehensive transport service information platform. This new smart system will be developed to facilitate issues of road conditions, the availability of car parking, and route navigation in the area. A central traffic information platform is also under development as a mobile app, which is expected to be operational in the coming years (estimated between 2020 and 2025). Since the project is being developed at a comparatively late stage (particularly in comparison to other main CBD areas of Shanghai), it can also learn lessons from similar cases of development in the region. Its geographical advantage nurtures its economic growth, and its policy advantage is expected to model the project as the 'demonstration area of low-carbon development and smart city' (Shanghai Hongqiao CBD 2016b). According to local government reports, the documents issued by the Ministry of Finance and Ministry of Commerce, emphasise the HQCBD zone as an integrated pilot model area for the development of modern service industries. These include promoting a wide range of 'management system and mechanical innovation of the modern service industry and aim to create the best systems, mechanisms and policy environment for the service industry' (ibid.). Finally, the case represents a unique approach to combine green/eco strategies and low carbon initiatives with a more recently developed smart development strategies; a hybrid model that can be considered as one of the potential future directions in sustainable urbanism (Cheshmehzangi 2017).

6.8 The Beijing Olympic Village, Beijing

Type: Eco-Residential Compound

Project Overview

The Beijing Olympic Village (BOV) is a residential complex within the Beijing Olympic Park, providing accommodation for contesting athletes, coaches and other accompanying officials. It was opened in July of 2008, in preparation for the 2008 Summer Olympics in Beijing, China. The development of the BOV follows Article 42 of the new Olympic Charter, which aims to situate competitors, team officials and team personnel in one central residential location by providing an Olympic village for a stipulated time period as determined by the executive board of the International Olympic Committee (International Olympic Committee (IOC) 2003). The BOV project involved a strategic partnership between key parties, including: the Beijing municipal government, the Beijing Ministry of Science and Technology, and the US Department of Energy, with representations from Japan and Germany. Guoao Investment and Development Company served as the developer, while Beijing Tainhong Yunfang Architecture Group worked on the design of the project (Beijing Organizing Committee for the Olympic Games 2008).

The village is situated towards the northern fringe of Beijing's north–south axis, and bounded by the Olympic forest park to the north and the famous 'B Nest' stadium to the south. It covers an area of approximately 660,000 m^2, with about 40% of its land use devoted to green spaces. The project includes 42 high-rise residential buildings (consisting of 20 nine-storey buildings, and 22 six-storey buildings), several public buildings, and separate facilities for leisure, entertainment and medical services. It is segmented into different areas, including: the athletes' village able to accommodate about 16,000 contesting athletes; media village (consisting of Green Homeland media village and the Huiyuan apartments media village) able to accommodate up to 7000 media staff; and two hotels able to accommodate around 800 guests (Pasternack 2008; Jia et al. 2012).

Eco-Initiatives and Green Features

The BOV was developed, by taking into consideration modern urban sustainability practices, including: The utilisation of renewable energy technologies, water reclamation, green building materials, ecological landscapes, and energy-saving buildings. It has been described as the world's largest green building complex, employing a mix of both high and low-tech solutions in reducing energy usage and increasing energy efficiency (Pasternack 2008). Key eco initiatives within the BOV project include (Beijing Organising Committee for the Olympic Games 2008; Pasternack 2008; Jia et al. 2012; Lassalle 2012; China Net 2008):

Water efficiency: Water conservation measures included the use of underground wells for irrigation and stormwater management, using permeable bricks in the pavements to form a rainwater recovery system, and applying micro-irrigation technology in order to water the village's green spaces. In addition, a biological sewage treatment plant with greenhouse structures was used to purify wastewater, thereby recycling between 200 and 300 tonnes of water daily to watering the landscape in the BOV.

Energy efficiency: the BOV employed a heat exchange system, which gathers up to 22 MWh of thermal energy from the Sun and recycled water. The system taps energy from the Qinghe Sewage treatment plant and upgrades it through a heat pump for cooling and heating during summer and winter, respectively. It saved 40% of the energy in comparison with a conventional air conditioner. The garage space for 2800 vehicles utilised T-8 fluorescent tubes with electronic ballasts and automatic controls. Furthermore, 6000 m^2 of solar energy collecting tubes provided bathwater to the 16,000 users during the games and 2000 users after the games, saving an estimated 5 GWh/year worth of electricity. Other new green technologies employed include solar heating, solar hot water, solar thermoelectric co-generation, optical pipes, LED lighting technologies and double-layered LOW emissivity glass windows. With these technologies, the BOV buildings used only 3.3% of energy consumed in modern buildings and annually saved about 147 MWh of energy. In addition, geothermal energy, using ground source heat pumps, radiant floors, desiccant cooling with active solar regeneration and seasonal thermal-storage system, helped to meet heating, ventilation and cooling demands.

Urban Green Spaces and Gardens: As stated in the above section, about 40% of the land uses in the BOV were devoted to green spaces. Also building roofs were planted with vegetation; with native plants making up 90% of the vegetation. This was one of the main key performance indicators of the project.

Case Study Reflections
Prior to the commencement of the 2008 Beijing Summer Olympic Games, plans had been made for the BOV to be remodelled and sold off as private apartments. An estimated 70% of the apartments were sold off

prior to the games, at 16,800 RMB per square meter in December of 2006 while the remaining 30% were sold off after the games (Xinhua 2008). The BOV functioned as the Paralympic village for the Paralympic games immediately after the Beijing 2008 Summer Olympics, as such remodelling work in the village and subsequent occupation by the new owners began after both games were concluded. Therefore, the project has taken on board a flexible approach to reuse the facilities and remodel the operations and services.

There are two key points that we can draw from the experience of the BOV: Firstly selling off the apartments in the BOV was an effective strategy to ensure the posterity and sustainability of its facilities, while making an appreciable return on investments for the developers in the post-Beijing 2008 Summer Olympics Games period. Examples abound from previous Olympic Games (Amatulli 2016; Bradley 2016), where such facilities are often left to decay due to the lack of appropriate plans for their sustenance. And secondly, hosting the Beijing 2008 Summer Olympics, in a city that was largely seen as one of the world's most badly polluted, was China's opportunity to demonstrate its commitment and passion for reducing the city's and the entire country's pollution and carbon footprint (Lassalle 2012). Seeing that a lot of the athletes, officials and global media outlets were residents in BOV, the project was strategic to demonstrate China's commitment to global sustainability.

6.9 The Grand MOMA, Beijing

Type: Eco-Residential Compound

Project Overview
The Beijing Grand MOMA is a linked hybrid set of eight high-end, high-rise apartment buildings in Dongzhimen Beijing, designed by the renowned American architect Steve Holl, which was privately developed by Modern Green Development Co., Ltd. The project sits on a land area of 220,000 m^2 and consists of mostly residential apartments, limited office spaces, stores, a movie theatre, library, restaurants, recreational facilities, a Montessori and kindergarten school. It is a mixed-use development, with residential use as its primary land use. Situated in close proximity to the eastern fringe of Beijing's second ring road and adjacent to the famous embassy neighbourhood, the Grand MOMA promises residents a safe and

conducive living condition while adhering to strict eco-friendly environmental practices (China Daily 2008; Steven Holl 2009).

Eco-Initiatives and Green Features
Environmental sustainability was a crucial requirement in the planning and eventual development of the Grand MOMA. The complex was developed for minimal energy consumption, energy conservation and pollution control. We briefly discuss three key eco-friendly environmental practices embarked upon within the Grand MOMA (Pasternack 2006):

Building Energy Management and Control: The Grand MOMA was designed to utilise passive energy sources. This is based in part on the passive house technology, which stipulates key requirements in architectural, technological and ecological planning. Water stored in a number of 100 metre-deep caissons beneath the underground parking lot of the complex was used to heat and cool the buildings, thereby reducing reliance on conventional heating, ventilation and cooling (HVAC) systems. This was made possible as a result of reliance on the geothermal energy heating principle, where underground stored temperatures can be leveraged upon to heat up and/or cool the water stored in the caissons. Heated and/or cooled water from the caissons is funnelled through pipes embedded within the floor slabs, helping to maintain temperatures within buildings at a range of 20–26°C. In addition, heat conservation within the buildings is achieved by equipping the windows with argon gas-filled double-paned glass, which also serves as shield from ultraviolet rays from sunlight.

Wastewater Recycling: The grand MOMA leverages on membrane biology technology for its wastewater treatment. It is reported that an estimated 58% of its wastewater collected from bathing and cooking activities of its residents are passed through a bioreactor and other sterilising devices, for recycling. The recycled water is then channelled back into the building for domestic usage and also watering and maintaining the landscape of the environment.

Pollution Control: Given Beijing's reported high rate of pollution, the Grand MOMA applies a proactive pollution control mechanism. The air coming into the buildings are first captured and screened to remove air particles. The air is then sterilised for contamination, after which it is regulated to the appropriate temperature. The clean air is then released into the rooms within the buildings, to create a safe indoor air quality environment.

Case Study Reflections
The linked hybrid-like structure of the Grand MOMA represents a very unique architectural creation, with its entire structure fused together to create an interactive space for its residents. In addition to its beautiful aesthetic qualities, the Grand MOMA prides itself on its ability to produce about four times the heating and cooling ability of conventional and traditional HVAC systems. The project continues to inspire architectural discourse, particularly with its brilliant adoption of geothermal technology to heat and cool the buildings within the compound. This represents a positive leap forward for eco-development; it helps to reduce reliance on traditional HVAC systems, thereby helping to curb the carbon footprints from such systems.

6.10 Conclusions

This chapter has presented eight meso-level case studies, categorised into four types of development: Eco-Industrial Park, Eco-Village, Eco-Business Park, and Eco-Residential Compound. These case studies represent a variety of sustainability practice at the neighbourhood/community level in China. All of them have adopted a development mode of mixed land use. As discussed in Sect. 2.3.6, the main focus at meso level is on the key aspect of 'patterns' that include neighbourhood forms, neighbourhood layout and related spatial qualities. Neighbourhood patterns should be able to facilitate walking and cycling, and promote interaction between people in public spaces. In this regard, the model of mixed-use development has a significant impact on neighbourhood performance as a whole. Technically, people living in a mixed-use neighbourhood would have a better chance of finding a job nearby, and a better accessibility to various facilities by walking and cycling. This reduces car dependency and changes people's travel behaviour accordingly. A survey was conducted by Qin and Qi (2015, pp. 113–115), which investigated the environmental performance of ten neighbourhoods in Beijing. This research indicates that neighbourhoods built after 2000 tend to have higher carbon emissions while older neighbourhoods would have lower emissions, some of which are due to the new added car-oriented housing development pattern (e.g., the increased incidences of gated communities in China). This is partly attributed to that the neighbourhoods built before 2000 are higher land use mixed with more facilities such as restaurants, parks, cinemas and public transport means in a walking distance, as well as more employment opportunities. Most of these are due to many factories or office buildings that are built in the nearby area (Qin and Qi 2015, p. 113). This also reflects a fact that Chinese cities are sprawling—but perhaps different in terms of pattern to the western models—and that many neighbourhoods built in newly developed areas lack public facilities and employment opportunities. More importantly, most of the daily commuting activities, as minimal as daily household shopping, place substantial reliance on private cars.

Physically, neighbourhood is a combination of multiple building plots, and public open spaces such as roads and parks. As an 'organised area of land', neighbourhood is universally subject to laws, planning regulations and urban/rural masterplans. Neighbourhood is a nexus between individual buildings and the broader urban context. However, in our view, the meso-scale built environment is still largely overlooked in the current eco-development practice in China. We have not found a case that directly aims to achieve zero-energy or zero-carbon performance, akin to the BedZED project in London (see Chap. 3). To date, only nine neighbourhood projects have been certified by LEED-ND in China (Green Building Map website 2017) and China's rating system for neighbourhoods is not scheduled to be available until July 2017 (and then released in October 2017). This local rating system is in place 11 years after the establishment of the building-level rating system. We also conclude that achieving a zero-energy neighbourhood is more complicated in the Chinese context due to the higher building density and building energy load (i.e., particularly, in comparison to the European models). This surely requires a concerted effort between buildings and spaces. For example, this is possible by utilising the open spaces for renewable generation, meaning that the power generated from open spaces can be fed back into building operations. Technologies, such as combined cool, heating and power (CCHP), ground source heat pumps, district cooling and heating systems, and smart grid are more suitable to apply at the meso level. An integrated set of strategies addressing the mutual impact between buildings and spaces is necessary to develop zero-energy neighbourhood in the future. The next chapter will discuss a few case studies at the micro level and further explore the strategies, technologies and evaluation tools in use in China. Some of these forthcoming examples are currently known as the main eco-/green building projects in China, and some, although outdated, provide us with the background of micro-level development and strategies in China.

REFERENCES

Cheshmehzangi, A. (2017) Sustainable Development Directions in China: Smart, Eco, or Both?, in the 2017 International Symposium on Sustainable Smart Eco-city Planning and Development, 11–13 Oct 2017 Cardiff, UK.

de Jong, M., et al. (2013) Developing robust organizational frameworks for Sino-foreign eco-cities: comparing Sino-Dutch Shenzhen Low Carbon City with other initiatives. *Journal of Cleaner Production*, 57, pp. 209–220.

de Vries, J. and Dai, G. (2015) Place making in Shanghai Hongqiao Business District: An Institutional Capacity Perspective, Urban Policy and Research, pp. 1–17.

Jia, H., et al. (2012) Planning of LID-BIMPS for Urban runoff control: The case of Beijing Olympic Village, *Separation and Purification Technology*, 84: 112–119.

Ghiglione, S. and Larbi, M. (2015) Eco-cities in China: Ecological Reality or Political Nightmare. *Journal of Management and Sustainability*, Vol. 5, No. 1.

Gubeikou Township Government (2008) Introduction of Implementing green and low carbon development in Gubeikou Township, Journal of Development of Small Cities and Towns, pp. 31–34.

Ling, Y. et al. (2013) Overview of Green Building label in China. Renewable Energy, 53, pp. 220–229.

Lassalle, J. L. (2012) Global Sustainability perspective. Eco-cities: Inspiration and Aspiration.

Pereira, A. A. (2003) State collaboration and development strategies in China: the case of the China-Singapore Suzhou Industrial Park. London; New York: RoutledgeCurzon, p. 47. (Call no.: RSING 338.951136 PER). Available from: http://eservice.nlb.gov.sg/item_holding_s.aspx?bid=11929058.

Qin, B., and Qi, B. (2015) Beijing, in Toward Low Carbon Cities in China: urban form and greenhouse gas emissions, edited by Han, Green and Wang, pp. 101–122, Routledge, New York.

Roseland, M. and Spiliotopoulou, M. (2017) Sustainable Community Planning and Development, Encyclopedia of Sustainable Technologies edited by Martin Abraham, Elsevier Oxford.

Singbridge 2016 'Sharing Singapore's Successful Development Experience', in Wong, J. and Lye, L.F. (eds) Singapore–China relations 50 Years. World Scientific: 153–176.

Yang, W., et al. (2017) A state of the art review on interactions between energy performance and indoor environment quality in Passive House buildings. *Renewable and Sustainable Energy Reviews*, 72, pp. 1303–1319.

6.10.1 Websites

Amatulli, J. (2016) Huffington Post: These Eerie Photos Show the Ghosts of Olympics Past (Internet). Available at: http://www.huffingtonpost.com/entry/these-eerie-photos-show-the-ghosts-of-olympics-past_us_57ab407fe4b0ba7ed23e5988.

Amy, F. (2013) Central Business District at Shanghai Hongqiao Airport by MVRDV with Aedas (Internet). Available at: https://www.dezeen.com/2013/04/24/central-business-district-at-shanghai-hongqiao-airport-by-mvrdv-and-aedas/.

Ascendas OneHub GKC (2016) Project website (Internet). Available at: http://www.onehubgkc.com/en/.

Beijing Miyun (2017) About Miyun (Internet). Available at: http://www.ebeijing.gov.cn/feature_2/BeijingMiyun/.

Bradley, A. (2016) Olympic villages past and present (Internet). Available at: https://www.realestate.com.au/news/olympic-villages-past-and-present/.

Beijing Organizing Committee for the Olympic Games (BOCOG) (2008) A preview of the Olympic village (Internet). Available at: http://www.china.com.cn/zhibo/2008-03/05/content_11638285.htm?show=t.

Beijing Tourism (2016) Gubeikou Great Wall (Internet). Available at: http://r.visitbeijing.com.cn/html/2013/Attractions_0814/186.html.

China Daily (2008) Ecological residential community rising in Beijing (Internet). Available at: http://www.china.org.cn/english/environment/241185.htm.

China Net (2010) Some Thoughts on Urban Audit Work in Expo City Best Practice Area (Internet). Available at: http://lianghui.china.com.cn/expo/2010-10/19/content_21151887.htm.

China Net (2008) The Olympic Village fast facts (Internet). Available at: http://www.china.org.cn/2008-03/05/content_11646502.htm.

China–Singapore Suzhou industrial park (SIP) website, Sino-Singapore Corporation (Internet). Available at: http://www.sipac.gov.cn/english/InvestmentGuide/SinoSingaporeCooperation/201107/t20110704_102985.htm.

China–Singapore Suzhou industrial park, 2010–2017. Infrastructure and Ecology (Internet). Available at: http://www.sipac.gov.cn/english/categoryreport/InfrastructureAndEcology/.

EU SME Centre website (2011–2017) About Suzhou industrial park (Internet). Available at: http://www.eusmecentre.org.cn/article/about-suzhou-industrial-park.

Fenghua Government website (2017) Overview of Fenghua (Internet). Available at: http://www.fh.gov.cn/zjfh/fhgl/.

German Energy Center & College website. German Enterprise Center Qingdao (Internet). Available at: www.german-energy-center.com/index.php?id=148&l=1.

Green Building Map (2017) (绿色建筑地图) Available at: http://www.gbmap.org/.

Guangzhou Knowledge City Project (2016) Overview (Internet). Available at: http://www.onehubgkc.com/en/gkc%20project/overview/.
Gubeikou Government website. Gubeikou: Basic situation of Gubeikou town in Miyun district of Beijing (Internet). Available at: http://gbk.bjmy.gov.cn/index.php?m=List&a=index&cid=1.
Han, G., (2010) Tengtou Village: China's eco-friendly example (Internet). Available at: http://www.china.org.cn/travel/expo2010shanghai/2010-08/06/content_20654703.htm.
Holl, S. (2009) Linked Hybrid: Description (Internet). Available at: http://www.stevenholl.com/zh/projects/beijing-linked-hybrid.
International Olympic Committee (IOC) (2003) Olympic Charter, ISBN 92-9149-001-6. Available at: http://www.joc.or.jp/olympism/charter/pdf/olympiccharter200300e.pdf. Accessed 22 September 2017.
Pasternack, A. (2006) Beijing's eco-friendly architecture (Internet). Available at: https://www.chinadialogue.net/article/show/single/en/635-Beijing-s-eco-friendly-architecture.
Pasternack, A (2008) Beijing's Olympic Village is Worlds' Largest Green Neighborhood (Internet). Available at: https://www.treehugger.com/corporate-responsibility/beijings-olympic-village-is-worlds-largest-green-neighborhood.html.
Ramanujam, M. (2015) Building the Shanghai of tomorrow: Hongqiao CBD's low-carbon transportation hub (Internet). Available at: https://www.usgbc.org/articles/building-shanghai-tomorrow-hongqiao-cbds-low-carbon-transportation-hub.
Sino-German EcoPark website. Project website (Internet). Available at: http://www.sgep-qd.de/ecopark/.
Sino-Singapore Guangzhou Knowledge City (2017) Press Release: Sino-Singapore Guangzhou Knowledge City achieves two significant milestones (Internet). Available at: http://www.ascendas-singbridge.com/~/media/asb-new/press%20releases/2017/ssgkc-groundbreaking.ashx.
Siemens (2017) Qingdao's Sino-German Ecopark, China (Internet). Available at: https://www.siemens.com/global/en/home/products/buildings/references/sino-german-ecopark.html.
Shanghai Hongqiao Central Business District (2016) Project website (Internet). Available at: http://en.shhqcbd.gov.cn/2016-03/30/.
Shanghai Hongqiao CBD (2016) Why Hongqiao CBD?; Available at: http://en.shhqcbd.gov.cn/2016-03/31/c_50614.htm.
Sim, C. (2015) National library board Singapore, 2015. China–Singapore Suzhou industrial park (Internet). Available at: http://eresources.nlb.gov.sg/infopedia/articles/SIP_2015-02-17_101058.html.
Souza, C.-M. D. (2009) Bold vision, vibrant chords: Sino-Singaporean Suzhou industrial park. Singapore: Keppel Corporation Limited, pp. 15–16. (Call no.:

RSING 338.951136 DE). Available at: http://eservice.nlb.gov.sg/item_holding_s.aspx?bid=13201841.
Sustainable Seattle website. Available at: http://www.sustainableseattle.org/.
Tang, Z. (2016) 28 Companies sign up for Shanghai Hongqiao CBD core zone (Internet). Available at: http://www.chinadaily.com.cn/business/2016-03/08/content_23786054.htm.
Thomas, L. (2012) Shanghai focus turns to Hongqiao CBD (Internet). Available at: http://www.scmp.com/property/hong-kong-china/article/1068049/shanghai-focus-turns-hongqiao-cbd.
Ti, Z. (2015) Guangzhou Knowledge City starts to take shape (Internet). Available at: http://www.chinadaily.com.cn/regional/2015-07/04/content_21310977.htm.
Top China Travel (2017) Tengtou Village (Internet). Available at: https://www.topchinatravel.com/china-attractions/tengtou-village.htm.
Xinhua (2008) Beijing to sell last 400 apartments in Olympic Village (Internet). Available at: www.chinadaily.com.cn/olympics/2008-09/12/content_7023975.htm.
Xinhua (2017) Feature: China-Germany economic cooperation blossoms (Internet). Available at: http://news.xinhuanet.com/english/2017-06/01/c_136331953.htm.

CHAPTER 7

Micro Level: Green Building Cases in China

7.1 Introduction

7.1.1 Current Green Building Practice in China

In China, the concept of 'Green Building' was originated in the use of building using energy-saving methods. The Chinese central government began to take measures to reduce energy consumption in building as early as the 1980s. In 1986, the Civil Building Energy Saving Standard (JGJ26-1986) was implemented and an energy-saving target of 30% (compared with the buildings built in 1980) was set for all new construction projects. This requirement has been increased to 50% in 1995, and then to 75% in 2016. In 2005, the Public Building Energy Saving Standard (GB50189-2005) was published, which commits the government to making an effort to reduce energy consumption in public buildings such as schools, hospitals, shopping centres and office buildings. Along with the building energy saving development, the concept of Green Building was introduced and the central government has recognised the need for an assessment system. Consequently, a national three-star Green Building Evaluation System (GBES) was launched in 2006. This particular rating system has two types of certification—Design Certification and Operation Certification. While Design Certification can be applied prior to building completion, Operation Certification can be only attained after the building has been completed and has been operating for at least one year.

© The Author(s) 2018
W. Deng, A. Cheshmehzangi, *Eco-development in China*,
Palgrave Series in Asia and Pacific Studies,
https://doi.org/10.1007/978-981-10-8345-7_7

In 2007, the MOHURD issued the Green Building Evaluation Technical Rules to direct the assessment process. GBES has been updated in 2014 (ESGB-50378-2014) that introduced a new assessment structure and is expanded to address issues related to the construction process.

In addition to these building design codes and evaluation standards, China has also taken measures to encourage property developers or investors to certify their buildings. In order to achieve this, the government provides additional financial incentives in the form of government subsidies, accelerates certification processes, and offers technical and marketing assistance. These certified buildings normally have higher selling or rental prices, high occupancy rates and long-term energy and cost savings. Some local governments have made it compulsory to attain GBES certification. For example, the Commission of Construction of Shandong Province announced in 2014 that all public buildings, such as schools, libraries, hospitals, should meet at least the 2-star design criteria of GBES-2014, and that residential projects should meet at least the 1-star design criteria.

In recent years, a nascent green building market has been emerging in major cities such as Beijing and Shanghai. However, there is considerable regional variation in the level of development of green buildings in China. Most green building projects are located in the Beijing–Tianjin strip, the Yangzi Delta and the Pearl River Delta, the three regions who have experienced the most rapid rates of economic growth (Ye et al. 2013; Ling and Ye 2012). For instance, Geng et al. (2012) examined 42 three-star buildings certified between 2008 and 2010. They identified all these projects as public buildings, most of which were governmentally-funded demonstration projects. This implies that the private sector has insufficient incentives to pursue such a certification. However, this situation has changed in the last few years as green building certification is becoming a compulsory requirement in some cities.

Cost has been shown to be a major factor affecting the development of green buildings. Ling and Ye (2012) have investigated 55 green projects in China between 2008 and 2012 and found that the average incremented costs are 16, 35 and 68 RMB per square metre for residential buildings that were certified for one-star, two-star and three-star ratings, respectively; and 29, 136 and 163 for public buildings. Other studies suggest a higher incremental cost for green building development. Through an examination of 39 green buildings between 2008 and 2010, Sun and Shao (2010) propose an incremented cost of 63, 131 and 219 RMB for residential building certifications and 45, 176 and 320 RMB for public building certifications. The divergences are attributed to the scope of 'incremented costs' between a green building and a conventional building for benchmarking. These two studies found that cost variation is also related to the geographical regions investigated. The studies also indicated that the incremented cost for public building is higher than residential building. Pursuing the three-star certification may incur significant cost given the national subsidy is around 80 RMB per square meter.

Researchers have also found that green building developers may enjoy a price premium in the current Chinese property market. Zheng et al. (2012) found that property developers in Beijing have been increasingly promoting buildings' 'greenness' in order to gain a price premium. Their research showed the claim of achieving green building can bring 17.7% price premium for the developers in Beijing at the pre-sale stage.

As the largest building construction market in the world, it is commonly believed that China will become a leader in the global green building industry. In addition to incentive policies, the improvement of rating systems and the increase in people's awareness, it is expected that the green building market will continue to grow in China. Baoxing Qiu, the former vice minister of the MOHURD, estimated that China's domestic green building market is already worth US$213 billion. And it will continue to increase in future decades.

A horizon scanning of current green building practice leads to a classification of current green building practice in China, which tries to capture the 'implementation focus' of a green building project—that is, the key modes by which green building visions are (to be) realised. This includes:

- Passive design-oriented: improving building performance through passive approaches.
- Certification oriented: the purpose is to satisfy the requirements from a specific green building rating system.

This categorisation should not be understood rigidly, as typically a green building project can be expected to combine both of the implementation modes. In particular, the passive design approach is usually considered prior to active approaches. It is nevertheless useful to pinpoint key features of green building development and the key actors involved, reflecting a bias towards the consideration of selecting green strategies at the building level.

7.1.2 Passive Design

Passive design and passive solar design refer to systems that take advantage of climatic conditions, building sites, and materials. These are considered to be the central means to minimise energy use, while maintaining the comfort of occupants. They are, therefore, strongly related to the environment in which the passive design buildings are located. Passive approaches have a long history and are found in building design in any region with the consideration of vernacular architecture. Such design techniques have been practiced for thousands of years. Passive design buildings are intended to be in harmony with the natural environment, and aim to create a comfortable indoor environment by taking advantage of local potential in order to eventually achieve the minimisation of total energy use. The use of passive approaches also aims to minimise reliance on the capacity of technological systems (e.g., active systems). In a mild temperature, it is possible to avoid the use of fuel energy; however, active heating and cooling may be required in more rigid climates. Adequate consideration on passive solutions can achieve the maximum reduction in energy and the gap is covered by active measures such as solar PV if the design target is to achieve zero operational energy consumption. A more rigorous concept is the so-called 'zero energy life cycle', which is principally focused on the building's lifecycle. This requires the greater use of renewable technologies generated to offset the embodied energy. In this chapter, we will explore three passive design case studies—CSET, Bruck and SIEEB, which are Chinese examples of implementing such passive approaches. In all three cases, the emphasis is initially on the integration of environmental features such as sun, wind and other climatic elements.

7.1.3 Design for Green Building Certification

Current building environmental rating systems have enjoyed considerable success and are often used by the government to provide financial or policy incentives. Cole (2010) identifies a number of factors that have collectively generated the success of such assessment methods:

- The prior absence of any means to both discuss and evaluate building performance in a comprehensive way;
- The limited number of performance criteria present a complex set of issues in a manageable form;

- Public sector building agencies have used the building rating system as a means of demonstrating their environmental commitments, and have issued an industry expectation of what constitutes 'green' building design and construction;
- Manufacturers of 'green' building materials have been given the opportunity to make direct and indirect associations with the relevant performance criteria.

It is important to note that the use of this type of building assessment system is not solely for labelling purposes and marketing promotion. In many cases, they have been utilised as planning and design tools because they present a set of organised environmental criteria. Accordingly, many design teams have chosen a specific green building certification at the very beginning and have been primarily focused on achieving the certification criteria throughout the design process.

In China, a number of green building rating systems are being used in the market. In addition to the national green building evaluation system (GBES), which is supported by a government incentive mechanism, some international tools such as LEED, BREEAM, Green Mark and WELL are also used to certify building performance. As mentioned earlier, some provinces and cities have placed a compulsory requirement on achieving a certain GBES rating as a prerequisite for construction approval. Accordingly, designing for green building certification has become a significant factor in decision making along with other conventional factor such as cost, aesthetics and functionality.

In this chapter, we explore four cases of certified buildings—the Hangzhou Science and Technology Museum, the Shanghai Tower, the Binhai Primary School in Tianjin and Nanhai@COOL in Shenzhen. These building projects represent some of the current examples of micro scale, focusing on obtaining a green building certificate.

7.2 Center for Sustainable Energy Technologies (CSET), Ningbo, Zhejiang

Type: Passive Design Oriented

Project Description
The Centre for Sustainable Energy Technologies (herein described as CSET) is recognised as the first zero-carbon building in China. Located at the University of Nottingham Ningbo China (UNNC) in the City of

Ningbo, the CSET is an iconic energy conservation building which relies mostly on renewable energy sources and the adoption of energy-efficient design practices to manage the building's energy requirements. The CSET serves mainly as a research centre focused on renewable energy systems for both domestic and commercial building developments and for furthering research on issues related to urban sustainability, with spaces for research laboratories, offices, seminar rooms and research exhibition. Key components of the building include a thermal testing lab, laser particle analyser, climate chamber, façade testing facility, and wind tunnel facility, amongst others.

Designed by the Italian firm Mario Cucinella Architects, the CSET building design concept was inspired by traditional Chinese paper lanterns and fan-like shapes (Fig. 7.1). It has a spiral shape, and sits on a land area of 1200 m^2 with a total of six floors overlooking the UNNC campus environment. As a result of its symbolic design style and energy conservation abilities, the CSET building was awarded with a prestigious MIPIM award in 2009. This award was within the green building category in Cannes France and continues to serve as a reference point for green building development in China (CSET website n.d.; Mario Cucinella Architects website n.d.).

Eco-Initiatives and Green Features

The CSET building is uniquely designed to reduce its dependence on conventional energy sources. A blend of renewable energy sources, energy-efficient design practices and the adoption of geothermal energy principles helps to control its energy requirements, thereby reducing its carbon footprint. In the section below we briefly discuss key eco-friendly environmental practices within the CSET (CSET website n.d.; Floornature Architecture and Surface website n.d.; Baderia 2014):

Heating and Cooling: The CSET building employs geothermal energy heating and/or cooling principle, relying on a radiative heat exchange mechanism. During the winter season, a reversible ground source heat pump (GSHP) provides heating to the thermal slabs through pipes embedded within the concrete floor slabs. In most cases, however, indoor temperature within the CSET building can be maintained using preheated ventilation air (discussed under ventilation strategies below). In the summer, solar-based air-conditioning systems help to maintain indoor temperature. In addition, the GSHP is also used to support indoor temperature management during the summer season, particularly where low solar

MICRO LEVEL: GREEN BUILDING CASES IN CHINA 193

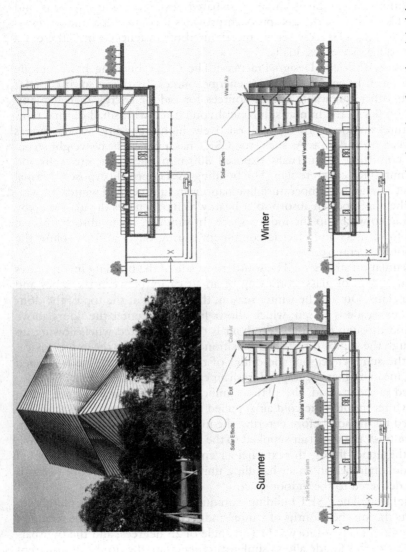

Fig. 7.1 Centre for Sustainable Energy Technologies (CSET), UNNC (Redrawn by the authors from the original documents provided by the CSET administrations)

radiation and high humidity are experienced. Given Ningbo's mild climatic conditions, a combination of solar-based air-conditioning systems and CSET's insulation design approach provides a comfortable indoor environment; the GSHP system is mostly used for instances where there is a need to balance excess loads.

Energy-Efficient Design Practices: The CSET building has a unique design as it leverages on passive energy sources. By doing so, it complements other renewable energy sources for indoor energy management. Ningbo lies in the humid subtropical zone often with small diurnal temperature variation (10°C) and relatively high humidity, which ranges between 65 and 95%. As such, the CSET has a typical heavyweight structure consisting of internally exposed 300 mm thick concrete walls and 400 mm thick concrete slab. The building uses internally exposed thermal mass to minimise temperature fluctuations in summer and winter. In winter, the envelope can absorb solar heat, which penetrates the glass thereby helping to warm up the indoor space. In summer, it can absorb internal heat from users and electric equipment during daytime for cooling the indoor temperature.

Ventilation strategies: The southern façade of the building incorporates a double skin of glass, consisting of an air cavity between the inner and outer skins. During the winter season, the opening at the top of the double skin façade is closed, which allows fresh air to enter the floors above ground through openings. Fresh air is then pre-heated when flowing up through the gap by direct solar radiation before entering the indoor space. For the summer season, the outlet of the double skin façade is opened fully, meaning that the trapped fresh air can be evacuated. A set of E-tubes, buried in the ground, also play a significant role in providing fresh air. In the winter months the cold air supplied to the semi-basement area is pre-heated by the soil before entering the indoor space. During the summer, by contrast, the warm air supplied to the building is pre-cooled by the soil. For the upper floors, the external air enters from the roof and is cooled and dehumidified in an air-handling unit located at the rooftop, which is then delivered to the indoor space.

Lighting: The CSET building's unique design helps to control lighting effects during the months of winter and summer. The southern façade of the building is tilted forward to an angle of 25 degrees, and the transparent part of this façade allows sunlight to penetrate the double transparent glass layers. This helps to allow sunlight into the building during wintertime given that the sun's altitude is low in the northern hemisphere. This design strategy also allows the level of sunlight to be controlled, meaning that it can deflect the amount of light allowed into the building during summertime when the sun is high in the sky. There is also a glazed strip on the eastern and western sides of the semi-basement area, to help maximise the penetration of natural light and avoid direct solar radiation. Moreover, PV solar systems are used to power the artificial lighting for the building.

Case Study Reflections
The CSET's strategic adoption of passive design strategies, and its reliance on renewable energy sources have greatly assisted in managing and controlling its energy requirements, thereby reducing its carbon footprint. Available reports show that the PV panels generated 42,009 kWh of energy, representing about 45% of the CSET's total energy consumption while the GSHP's, wind turbines and solar systems generated the remaining. The CSET boosts of saving 169 tonnes of carbon emissions, compared to conventional buildings of similar size and structure in the region. Finally, it should be stated not only that the CSET building is a remarkable architectural masterpiece, but that it also demonstrates some of the early eco-strategies of building design in China.

7.3   Bruck Residence, Huzhou, Zhejiang
Type: Passive Design Oriented
Project Description
The Bruck residence is a passive house energy-efficient building project located in Changxing County, Huzhou city in China's Zhejiang province. The project was developed by the Landsea Property Development Group in cooperation with Peter Ruge Architects, and was certified by the Passive House (Passivhaus) Institute in Germany. The unique nature of the Bruck residence project, apart from its obvious adoption of the passive house technology, is its location. The Bruck residence represents the first of its kind to be developed in a hot and humid climatic environment in the south of China, where climatic conditions are often temperate with brief and mild heating and extensive cooling periods, and humidity reaching levels above 80%. The project mainly employs principles from the passive house building technology in managing energy consumption, while creating a habitable indoor environment (Passive House Bruck website; Schoof 2015).

Completed in 2013, the Bruck residence sits on a land area of 2200 m² and consists of 46 apartments in a five-storey residential building. The developers mainly developed the Bruck residence as a demonstration project to showcase passive house development in hot and humid climatic environments, and also as part of their strategy to develop energy-efficient buildings in China. The building consists of a total of 36 single-room apartments reserved as living quarters for the staff of the Landsea property development group, six two-bedroom executive apartments, and four three-bedroom apartments reserved as demonstrations to prospective clients for the experience of living in a passive house (ibid.). Figure 7.2 shows the green systems employed in this building.

Eco-Initiatives and Green Features

The Bruck residence features unique design practices, which draw on principles from the passive house technology. In addition to strict adherence to building sustainability, durability and eventual maintenance, informed the choice of materials used in its design. Key initiatives embarked upon within the Bruck residence are discussed below (Archdaily 2014; Passive House Bruck website):

Building Insulation Design: The building was designed carefully, with strict consideration being given to the local climatic conditions in Huzhou. A concrete skeleton formed the core of the building, consisting of columns, beams and floors, with a heavily insulated envelope of 20 cm thick

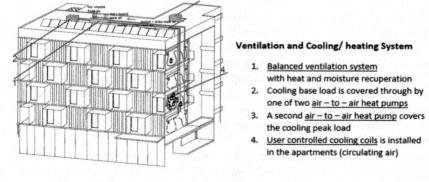

Fig. 7.2 Green systems employed in the Bruck Residence, Huzhou China (adapted and redrawn by the authors from the original system design)

batt mineral wool insulation. Particular attention was given to maintain air tightness within the building: for instance, both the roof and walls have U-Value of around 0.15 W/m²K and 0.20 W/m²K, respectively. Moreover, the interior walls are designed by using drywall in line with local guidelines for fire safety and acoustics.

Building Ventilation and Heat Recovery: A mechanical ventilation heat recovery unit (MVHR) is centrally pre-conditioned to 22°C and located on the roof of the Bruck residence. The MVHR unit involves a set of condensation rotary heat exchangers and a controlled rotating sorption: It records heat and humidity recovery rates of 75% and 60%, respectively. Dehumidification within the building is set to a maximum of 12 g/kg when necessary, by cooling and reheating the air supplied using the rotary heat exchanger and a groundwater-fed cooler. A central duct system in the energy recovery unit, located on the roof of the building, helps to remove exhaust air.

Building Exterior Shading Design: Huzhou's warm climatic condition called for measures to ensure appropriate measures into consideration for the enhancement of the building's shading ability, given that energy consumed for cooling in summertime is often usually high. Vertical terracotta-coloured shading elements are used to shield the building's envelope from solar insolation on the eastern, western and southern façades, with the same process applied to the hallways and stairways to the northern part of the façades. External sun louvres are also used to control the amount of sunlight allowed into the building. This is calculated according to simulation-based results, which are used to determine the size and positioning of the louvres for optimum shade efficiency.

Case Study Reflections

The development of the Bruck Residence in Huzhou, in a relatively warm climatic environment, goes against conventional practices in the development of passive house. Previous efforts to develop passive houses have often been concentrated in communities or cities with colder climatic environments, based on the assumption that passive houses are believed to work better in cold climates. The Bruck residence project directly challenged this assumption, and sets a good example in ensuring that sustainable design practices, and the use of passive house technology systems, can be undertaken for building projects situated in locations with similar climatic conditions. Available reports (Schoof 2015; Passive House Bruck website) suggest that the Bruck residence records a 95% energy saving in comparison to conventional buildings of similar size and capacity, and a 60% reduction in the amount of energy consumed when compared to other building projects undertaken by the same developer. Overall, the building sets an example of a high-performance passive house project in the warm climate zone of China.

7.4 Sino-Italian Ecological and Energy-Efficient Building (SIEEB), Beijing

Type: Passive Design Oriented

Project Description

The Sino-Italian Ecological and Energy Efficient Building project (SIEEB) is a strategic developmental cooperation between the Chinese Ministry of Science and Technology, and the Italian Ministry for the Environment and Territory. It represents a bilateral long-term partnership between both parties in terms of furthering research and development in key areas of energy and environmental sustainability, particularly with respect to addressing issues related to CO_2 emission in China's building sector. An integrated design approach was adopted in the development of the SIEEB building project, involving collaborations between key stakeholders from Tsinghua University, Politecnico di Milano, Mario Cucinella Architects and the China Architecture Design & Research Group (Batera, et al. 2005; Archdaily 2017, SIEEB website).

Located within the Beijing campus of Tsinghua University in China, the SIEEB occupies a total floor area of 20,000 m² and stands at a height of 40 metres (see Fig. 7.3 below). With 10 floors above the ground and 2 underground floors, the SIEEB mainly consists of spaces for offices, a 200-seat auditorium, and a Sino-Italian education training and research centre for environmental protection and energy conservation. The building's unique design involves two symmetrical wings with a centred atrium, which helps to ventilate the building, serves as space for interaction, and also as an aesthetic quality. In addition, extensions of ten platforms from the upper parts of the building are exposed to the Sun, serving as shields from sunlight and adding to the building's aesthetic quality (Architravel 2013).

Eco-Initiatives and Green Features

The design of the SIEEB integrates state-of-the-art technologies with sustainable building design practices, in consideration of Beijing's architectural and climatic environment. Both active and passive design strategies were adopted to better manage the energy requirement of the building, and ensure a comfortable indoor environment. Below, we briefly discuss two key eco-strategies in the design of the SIEEB (Batera et al. 2005; Archdaily 2017; Architravel 2013; Butera 2008):

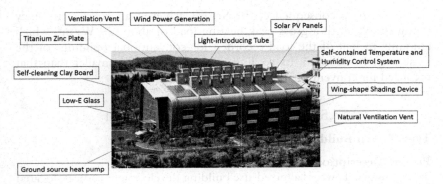

Fig. 7.3 Green systems employed in the Hangzhou Low Carbon Science and Technology Museum Building (images taken and adapted by the authors)

Building Insulation Design: The northern fringe of the building facing the cold winter winds is closed off and well insulated, while the southern part is left open and transparent. Incoming lighting and sunlight to the eastern and western parts of the building are controlled by a double-skinned glass façade, which acts as a filter to control the penetration of sunlight into office spaces. In addition, the exterior walls of the courtyard are covered in pivoting glass louvres with reflective coating, to help manage daylight and solar gain.

Building Ventilation and Thermal Comfort: Indoor thermal comfort within the building is enhanced through the help of a displacement ventilation system for distributing primary air, and a radiant ceiling system. Using the lightweight radiant ceiling system allows for lower air temperatures in the winter and a higher temperature in the summer. This helps to control the amount of electricity consumed for powering fans and pumps. Furthermore, a combined heat and power system (CHP), which forms the building's core energy system, supplies most of the energy required by using a series of gas motors coupled with electrical generators.

Case Study Reflections
The design and development of the SIEEB is unique in that it balances the goal of achieving energy efficiency in buildings with aesthetics. In particular, its integrated design process was pivotal in ensuring that every stakeholder was actively engaged in its developmental process. One aspect of this process involved extensive testing and simulation conducted on the building's shape, envelope, orientation and technological systems to ascertain the performance of the SIEEB in consideration of Beijing's climatic

condition. As a consequence, the SIEEB achieved an energy efficiency rating of 80% with a gross peak heating load of about 29 W/m^2, which amounts to half of the load recorded for conventional buildings of the same size. In addition, its cooling load is about 2.5% smaller than that recorded for conventional buildings at 80 W/m^2.

7.5 Shanghai Tower, Shanghai

Type: Green Building Certification Oriented

Project Description

The Shanghai Tower is a mixed-use building development project, located within the Lujiazui commercial district of the city. Situated on a floor area of 420,000 m^2, it rises to an impressive height of 632 metres, making it the second-tallest building in the world behind Dubai's 828 metre-tall Burj Khalifa. The tower consists of 128 floors above the ground and 5 underground floors. These floors are further segmented into nine vertical zones, consisting of between 12 and 15 floors per zone, each of which offers a mix of spaces for offices, shops, restaurants, hotels, an observation deck and a conference centre on the topmost floors.

The design of the Shanghai Tower is a unique blend of architectural aesthetics with sustainable building design practices. Specifically, one of the unique features of the project is its twisting form and curved façade. This was shown, using wind tunnel test, to lead to a 24% saving in structural wind loading when compared to rectangular buildings of a similar height (Gensler Design website; Rodriguez 2017). The key stakeholders involved in the project includes, among others, the Shanghai Tower and Construction Company Limited as the main developers, the American design firm Gensler, the structural engineering firm Thornton Tomasetti, and Mitsubishi Electric, who have provided the elevators. The building was certified a LEED Platinum in 2013.

Eco-Initiatives and Green Features

The Shanghai Tower prides itself as one of the world's most sustainable energy-efficient high-rise buildings, with an ambitious target of reducing its carbon footprint by 25,000 metric tonnes per year. In addition to its use of passive design practices, it also employs a mix of green energy sources to fulfil its energy requirements. Some of the key eco-strategies used in its design are briefly discussed below (Rodriguez 2017; Gensler Design website; Tetra Tech website n.d.):

Sustainable Building Design Practice: A number of design initiatives were undertaken in the Shanghai Tower. First, particular attention was paid to the physical form of the building with its proposed height, given Shanghai's susceptibility to typhoon-strength winds. As a result, a design depicting an asymmetrical form, rounded corners and a tapering profile was chosen after extensive tests in a wind tunnel. Secondly, two façades wrapped the building, serving both as an exterior atrium and as an insulating blanket, and thereby controlling the energy use for cooling and heating.

Renewable Energy Generation: The building employs a mix of energy sources to meet its requirement. Specifically, wind turbines, totalling about 270 and built into the building façades, are used to power exterior lighting for the building. In addition, a geothermal system and a 3.0 megawatt combined heat and power system (CHP) helps to manage the buildings energy requirements.

Rain and Waste Water Recycling: The Shanghai Tower adopts strategies for water conservation and consumption. Firstly, toilet systems and kitchen sinks that consume minimal amounts of water were installed in all apartments and necessary spaces within the building. Secondly, the building is equipped with systems to capturing and harvest rainwater and wastewater, which is then recycled and channelled for other uses.

Case Study Reflections
The Shanghai Tower demonstrates that, for a building of its size, eco-structures can be of mega sizes. Energy saving within the building is reported to be at about 21% reduced and the building's life cycle cost is put at a reduced 45% due to the use of sustainable building design practices. Although more use of renewables could have been demonstrated, particularly with regard to the façade, it is, notable that efforts and cognisance towards achieving energy efficiency, green spaces and socio-cultural gestures in the form of the sky lobby has been achieved.

7.6 Hangzhou Low-Carbon Science and Technology Museum, Hangzhou, Zhejiang

Type: Green Building Certification Oriented

Project Description
The Hangzhou low-carbon Science and Technology Museum (herein described as HLCSTM) is the world's first low-carbon-themed large-scale museum. Located in the city of Hangzhou, the capital of Zhejiang province China, it sits on a total land area of 4679 m² across four floors with

spaces for exhibition, research labs and offices. It also has an underground floor, used mostly for the storage of equipment. The China Energy Conservation and Environmental Protection Group developed the HLCSTM project in 2009, as a key part of a broader pilot initiative by the National Development and Reform Commission (NDRC) towards Hangzhou's transition to a low-carbon city.

It is primarily aimed at engaging and educating, mainly the residents of Hangzhou, with information on low-carbon technologies and to showcase advanced green building design concepts and technologies in China. The Chinese Academy of Building Research served as the main consultant of the project, in providing support for applicable green technologies and strategies. At the very beginning, the project team aimed to achieve the highest certification of the two green building rating systems—GBES three-star and LEED-Platinum. As a consequence, the HLCSTM was awarded two green building certifications; including GBES-three-3 star and LEED-Platinum, in 2009 (HLCSTM 2017; NCSC 2013; IVL 2016). Figure 7.3 shows the green systems employed in this building.

Eco-Initiatives and Green Features
The HLCSTM adheres to strict requirements of ecological sustainability, energy savings and carbon reduction. Some of the key eco-strategies used in its design are briefly discussed below [HLCSTM 2017]:

Environmentally Adaptive Envelope: The HLCSTM building project was developed using a steel frame structure. Titanium zinc panels are used in the north, south façade and roof while clay plates were applied to the west and east façades. All materials used are recyclable and reusable. The building envelope is well insulated and all the windows are double glazed with low-emissivity glass, which is highly transmitted but prevents long waves of heat from penetrating in the summer. The window size on the western and northern façades is minimised to reduce heat loss in winter. By doing these, energy simulation indicated that the load for heating and cooling has been decreased from 41.71 W/m^2 to 23.53 W/m^2.

Passive Ventilation System: This system functions mostly at night time during the summer season and at the transition point between summers, but is shut down in wintertime. Fresh air enters the semi-underground air ducts, where the water that is circulated through pipes buried in the ground cools it down. There are 14 vertical air ducts located in the north and south sides and 4 vertical ducts in the east and west sides of this building. Fresh air enters the indoor space through the 18 chimneys installed on the roof to enhance stack effect, the air moves into the atrium and is pulled outside from the roof chimneys and thus forms natural ventilation. The passive ventilation system is, however, turned off to maintain a lower interior temperature and humidity when the outdoor temperature and humidity is too high.

Shading System: The building leans to the south by 15°, which helps it to control direct solar radiation in summer due to a high solar altitude angle. Meanwhile, it allows sunlight to enter the building in winter when the sun position is lower. The external blinds on the south and north façades are controlled by an automatic system that is able to adjust the amplitude of the blinds. With help of the chimney effect, ventilation shutters open automatically to release redundant exhaust heat and reduce the indoor temperature under the passive ventilation mode. In rainy or windy weather, the shutter will close automatically. In addition, the shutter can open automatically to discharge smoke in a case of a fire event.

High-efficient Lighting System: A Solatube daylight system is used throughout the third floor. The dome is designed to filter out UV light. The light interception transmission device (LITD) has a high capacity of capturing sunlight. In addition, daylight is transported in the pipeline with an unparalleled reflection rate up to 99.7%. After passing through pipe and diffuser, daylight is able to lighten throughout the third floor in the daytime. Energy-saving fluorescent lamps are also installed in the office space and plant rooms. Voice-control sensors are installed in the corridors and stairs to further reduce energy consumption for lighting.

Ground Source Heat Pumps to Provide Heating and Cooling: Ground source heat pumps are installed to provide heating and cooling to this building. Water is recirculated as heat transfer media. Sixty-four U-shaped pipes are buried underground to a depth of 60 metres, where there is less variable temperature to increase the efficiency of heat exchange. Because of Hangzhou's relatively high humidity and its significant impact on indoor comfort, particularly in the plump rain seasons, four dehumidifying devices are installed to absorb and remove moisture using saline solution, thereby adjusting the indoor humidity level. Because of the high level of humidity in summer, these devices purify the fresh air and then send the dried air, by air ducts and terminal devices, to the rooms. One more advantage is that AC terminal units with low temperature are also protected from condensation.

Renewable energy: In total, 296 m^2 multi-crystal silicon PV panels are installed on the roof that can generate 40 kW power in a day, and 57 m^2

photoelectric glasses can produce additional 3 kW. On a typical day, solar power is able to meet 14% of daily electricity demands. In addition, two wind turbines can also generate 600 W power in a day that is stored in the hydrogen fuel cells as backup power.

Case Study Reflections

This project has demonstrated an excellent green building case with low-power, eco-beneficial, energy-saving and user-friendly features. The HLCSTM continues to serve as a key project for informing and educating the general populace, within Hangzhou and beyond, of the importance of low-carbon cities and applicable technologies. Apart from its adoption of eco-initiatives, its educational potentials on urban sustainability increasingly attracts high numbers of visitors, with the Ministry of Housing and Urban-Rural Development listing it as a national demonstration project for green building promotion in 2011. The HLCSTM building project boosts of an estimated energy saving ratio of 76.4%, with the annual completion less than 25% of the normal buildings of similar size and structure.

7.7 NanHai@COOL, Shenzhen, Guangdong

Type: Green Building Certification Oriented

Project Description

The NanHai@COOL project involved the retrofitting of an old industrial building into a modern building for office spaces. It is located in the Taizi Lu Nanshan District of Shenzhen city, and sits on a total land area of 25,023 m^2 (Fig. 7.4). The original industrial building on the site of the NanHai@COOL was a four-storey factory building with frame structure and no basement. The new structure has five floors: the basement and the first floors are garages, the second are fourth floors were reserved for office spaces, while the fifth floor mainly consists of activity and multi-functional rooms.

The project was designed by Tsinghua-Yuan Architectural Design Ltd Shenzhen, and it involved collaborations between key stakeholders, including: China Merchants Property Development Co. Ltd, the Building Energy Conservation Research Centre Tsinghua University, the Shenzhen Institute of Analysis and Test Center of Peking University, the Shenzhen Institute of Building Research Co. Ltd, the China Academy of Building

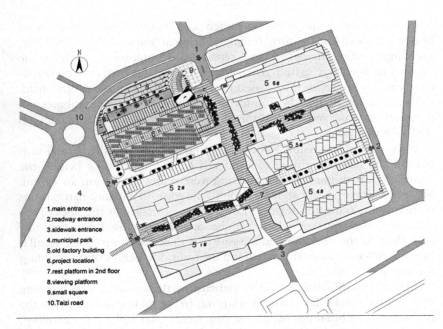

Fig. 7.4 The project floor plan (redrawn by the authors based on the original plan provided by Professor Yu Liu from the Northwest University in China)

Research Shenzhen Institute, and the Shenzhen Yuezhong Green Building Science and Technology Development Co. Ltd. The projects' adoption of green building initiatives was recognised, as it was awarded the highest three-star green building design label and three-star green building operation label in 2010 and 2013, respectively, amongst others (Wu 2010; Lin et al. 2010; Yang et al. 2009).

Eco-Initiatives and Green Features
In this section, we give a brief discussion below of some of the key ecological initiatives embarked upon in retrofitting the old industrial building into a modern building for office spaces (Wu 2010; Lin et al. 2010; Yang et al. 2009; Wang et al. 2008; Liu et al. 2011):

Building Materials Recycle and Reuse: The project explored new buildings forms, based on the retention of the original environment. Special efforts were made on strengthening and utilising the original structures (e.g., structural joints were restrained and bonded with epoxy resin bonded carbon fibre cloth), retaining original walls as far as possible, recycling existing materials, and reusing old equipment (such as the original high-voltage switch-gear). Nearly 1000 cubic metres of construction wastes were crushed and used for the backfilling of the project field. Old equipment, such as the transformer, a high-voltage switch cabinet and power cables were reused, and this alone helped save nearly three million RMB for the project.

Natural Ventilation and Natural Lighting: To improve the natural ventilation and natural lighting of a industrial building of some height, the second to fourth floors of the building were removed to form an atrium. Six solar chimneys were added on top of the roofs to enhance natural ventilation in the transition seasons. Louvres were installed at the sides of the chimneys, the vents of which can be opened or closed automatically according to the prevailing conditions in different seasons. Such ventilation systems were estimated to extend the transition period of the building for nearly two months.

External Sun Shading: In the Shenzhen area, the annual air-conditioning season is long and solar radiation is strong. In the middle of June, even the north facade of buildings receives strong direct solar radiation. Therefore, sun shading is essential for to assist buildings in energy saving. In this project, photovoltaic panels were applied as sunshade for the roof of the atrium area; horizontal sunshades were applied on the south facade; fixed horizontal, vertical sunshade, nearby buildings and tall trees were used to provide shading for the east facade; adjustable sunshade and vertical greening wall were applied on the west facade; and ladder-type space greening wall was applied on the north facade. These systems work together to achieve an estimated reduction of about two-thirds of the cooling load of the building.

Renewable Energy Technology: Shenzhen's average annual total solar radiation is about 5000 MJ/m^2. This project has attempted to take full advantage of solar energy. The roof of the building was equipped with 82 m^2 of solar thermal panels, which can produce 100% of hot water for washing and shower. A ground source heat pump system was designed to provide a supplementary source of energy to ensure hot water is provided throughout the year.

Temperature and Humidity Independent Control (THIC) Air-conditioning System: This was the first large-scale application of the THIC system in south China. In this system, two sets of independent systems are used to control and adjust indoor temperature and humidity, respectively. These can easily meet the different requirements of indoor temperatures and humidity (conventional air conditioning system is difficult to meet the temperature and humidity parameters at the same time); therefore can avoid the indoor humidity to be too high or too low.

Non-traditional Water System and Artificial Wetland: A stormwater collection system and a greywater reclamation system were applied in this project. To achieve zero discharge of sewage, the reclaimed water is completely recycled and reused. The greywater reclamation system uses artificial wetland-processing technology which includes the following: first the shower drainage, washbasin drainage and other high-quality miscellaneous drainage are discharged to the No. 2 wetland for treatment and the reclaimed water is mainly used as supplement for landscape water irrigation. Secondly, the toilet and canteen drainage are disposed to No. 1 wetland for preliminary treatment and the final reclaimed water is used for toilet flushing. About 25 m^3 of greywater is treated each day, meaning that around 5000 to 6000 m^3 of civil water is saved on an annual basis. Rainwater on the roofs are also collected and mainly permeated back to the ground to supplement underground water, some of which is also used for cooling tower water supplement, flushing toilets and cleaning. Theoretically, about 8000 m^3 of rainwater can be collected annually from the roofs, meaning that 2800 to 3200 m^3 of rainwater can be used if a figure of 35% to 38% of is reclaimed.

Case Study Reflections
With the rapid development and transformation occurring in China's economy in recent years, a large number of former factory or industrial buildings are being retrofitted or demolished. As was the case with NanHai@COOL project, such transformations present opportunities to adopt sustainable building design practices either in retrofitting old buildings or the development of new building projects. The NanHai@COOL project particularly serves as a model in this regard given its recognition and subsequent energy certification, for its strict adherence to urban sustainability and energy-efficient building design initiatives. Broadly speaking, this project will play an excellent role in demonstrating a range of green building design strategies and technologies and as a good example for the retrofitting of old factory buildings in China.

7.8 Binghai Xiaowai Primary School, Tianjin

Type: Green Building Certification Oriented

Project Description

Binhai Xiaowai Primary School is a low-energy consumption school building project, located within the Sino-Singapore Tianjin Eco-City in Tianjin China. Situated on a land area of 53,000 m², the project is a mixed-use educational building facility consisting of spaces for classrooms, laboratories, staff offices, libraries, student dining halls, recreational and sports facilities, amongst others. Due to constrained space requirement for the development of the project, an S-shaped design model was adopted for the 36-standard classroom primary school with spaces and facilities uniquely distributed along the S-shaped structure.

The project was designed by designers from the Beijing-based design studio, HHD_FUN, and is currently rated three-star in the national green building label (HHD_FUN 2008; HHD_FUN 2015).

Eco-Initiatives and Green Features

Given Tianjin's often-cold and relatively dry climatic conditions, the design and eventual development of Binhai Xiaowai Primary School followed strict adherence to requirements for environmental and building energy conservation. Below, we briefly discuss some of the key eco-initiatives embarked upon in Binhai Xiaowai Primary School (HHD_FUN 2015; Kwok 2015; Vyas 2016):

Efficient Envelope Structure: Keen attention was paid to the building structure with respect to the use of energy-saving materials. A double silver Low-E insulating glass fibre profile, with an air layer of 12 mm, was adopted for the outer window, resulting in an average heat transfer coefficient of 1.5 W/m²K. For the roofing, a composite rock wool insulation board with an average heat transfer coefficient of 0.21 W/m²K was adopted, while the external walls of the building was insulated with materials resulting in an average heat transfer of 0.21 W/m²K.

Building Shading and Lighting: The southeast and southwest façade of the teaching building was affixed with a wing-type adjustable outside shading system, which had no impact on the indoor natural lighting. This system automatically controls the sun louvre angle vis-à-vis the exposure from sunlight, by relying on weather information from a small weather station positioned on the rooftop of the building. In addition, a high atrium space with transparent rooftops was constructed in the corridor of the building to allow natural lighting and reduce energy consumption from providing conventional lighting systems.

Heat Recovery and Gas Source VRF Systems: The building was equipped with heat recovery units, located in its basement. A digital intelligent energy-saving heat recovery unit was used for exhaust heat recovery and to treat fresh air. These units were set to operate at optimal efficiency, with the recovery efficiency of the whole heat exchanger being not less than 60%. Furthermore, the building project employs a gas source variable refrigerant flow air-conditioning system. This system serves as additional sources of cooling during the summer season, and heating during the winter season with the COP (coefficient of performance) reported as 3.73.

Renewable Energy Use and Water Conservation: The building is also equipped with a decentralised solar-powered hot water system. The system leverages on thermal energy from the Sun to heat up and provide hot water for kitchens and shower rooms within the building. A small rainwater collection pool was constructed towards the northern side of the building, which channels rainwater collected from the roof for irrigational purposes. Additionally, an outdoor ecological science park was created to educate students on water conservation.

Case Study Reflections

The adoption of green building technologies helps to control building energy consumption, as well as provide occupants with a more comfortable indoor environment. This, however, impacts substantially on the cost of developing the building project. Available reports show that the cost of the Binhai Xiaowai Primary School building project increased by 377 RMB/m^2 as a result of its green building initiative, with the cost of the outer window swallowing up 45% of the increase, followed the solar-powered hot water system accounting for 16% of the total increase in costs. The project does, however, lead to tremendous energy cost saving over time, with annual energy-saving estimates placed at 54 million RMB and the static investment recovery period set at about 19 years.

7.9 Conclusions

This chapter has presented seven green building cases across the country. The categorisation of these cases is not rigid, as all of the cases considered adopted a combination of passive and active measures. Passive measures focus on maximising external environmental benefits while active measures aim to optimise building mechanical systems to provide a comfortable indoor environment. These measures often include utilising renewable energy, such as solar energy, wind energy and geothermal energy, improving daylight and energy-efficient illuminations, and the increasing efficiency of HVAC systems. They are well-established technologies and are also applied universally in many other countries in the world. Some projects have been certified by international green building evaluation systems, such as Shanghai Tower by LEED and Bruce Residence by the Passive House Standard in Germany. This reflects that international experience has been used extensively in the Chinese green building development.

However, on the other hand, it should be noticed that green building tools and technologies are also contextually based and it is not a right approach for China to merely replicate foreign tools/technologies without any adaptation or localisation. This is argued based on the following facts:

- Chinese cities usually have higher population and building density that may restrict the installation of renewable technologies such as solar PV and wind turbines, and the mutual effect between adjacent buildings should not be ignored;
- China is the third largest country and has diverse climatic conditions and geographic features. From the perspective of building energy design, it is generally divided in to five climate zones, from severe cold to tropical climate. These climate zones require contextually adapted tools and technologies and cannot be simply unified into one category;
- China's average building energy consumption is lower than in most developed countries; however, this is concurrent with low thermal comfort in buildings, particularly in the Hot Summer and Cold Winter (HSCW) Zone. This region has a low temperature during the winter time but no centralised heating system is provided. As a result of this situation, most people use self-standing heater devices or air conditioners during the cold—but relatively short—period of winter. In this respect, building technologies should be able to increase indoor comfort and reduce energy consumption in tandem.

Therefore, it is important to combine the contextual conditions and advanced tools and technologies being adopted in a reference building to develop green buildings locally. The next chapter will look at the past and current trends of eco-development at three spatial levels and give an overview of benefits from the concept of eco- and/or green from a multiscalar perspective. By doing so, we then focus, in particular, on the further discussion on the context of China and its past, current and future directions.

REFERENCES

Baderia N. (2014) The Role of Thermal Mass in Humid Subtropical Climate: thermal performance and energy demand of CSET Building, Ningbo. 30th International PLEA Conference 16–18 December 2014, CEPT University, Ahmedabad.

Batera, et al. (2005) The Sino-Italian Environment & Energy Building (SIEEB): A model for a new generation of sustainable buildings. International Conference on Passive and Low Energy Cooling for the Built Environment, May 2005, Santorini, Greece.

Butera F. (2008) The Sino-Italian Environment & Energy Building (SIEEB): A Model for a New Generation of Sustainable Buildings. In: Clini C., Musu I., Gullino, M. L (eds.) Sustainable Development and Environmental Management, Springer, Dordrecht.

Cole R.J. (2010) Environmental assessment: shifting scales, In Edward Ng (eds) *Designing high-density cities for social and environmental sustainability*, pp. 273–282, Earthscan London.

Geng Y., Dong, H., Xue B., Fu J. (2012) An overview of Chinese Green Building Standards, *Sustainable Development*, vol:20, issue:3, pp. 211–221.

Lin W., Dong J., Wu Y. (2010) Retrofitting practice of Naihai Ecool building in Shen Zhen [J]. Architectural Journal, 2010, (01):20–25.

Ling, S., and Ye Z. (2012) Study on the Economics of Green Buildings in China, China Sustainable Energy Programme (Serial No: G-1110 14964).

Liu S., Liu X., Jiang Y., Liu X., Li H., Chen X., Qiang B., and Yang H. (2011) Performance test and analyses of temperature and humidity independent control air conditioning system in an office building in Shenzhen [J]. Heating Ventilating & Air Conditioning, 2011, (01):55–59+8.

NCSC Energy Foundation (2013) National Centre for Climate Change Strategy and International Cooperation, Low carbon pilot project situation in China and its future research direction.

Sun, D.M., Shao, W.X. (2010) Statistic and Study on Incremental Cost of Green Building in China, *Eco-city and Green Building*, 2010–04 (in Chinese).

Wang S., Yang S., Zhu L., Wang J., (2008) Ecological function of courtyard – A case study on reconstruction of No. 3 plant building in Shekou [J]. Industrial Construction, 2008, (11):49–51.

Wu, X. (2010) To probe into the conversion of existing industrial buildings to creative industry park in Shenzhen: Analyzing and comparing OCT – LOFT and Nanhai E-Cool [J]. *Architectural Journal*, 2010, (S1):47–50.

Yang H., Liu S. and Liu X. (2009) Operating practice of temperature and humidity independent control air conditioning system in an office building in Shenzhen [J]. Heating Ventilating & Air Conditioning, 2009, (05):135–138.

Ye, L., Z. Cheng, Q. Wang, W. and Ren, F. (2013) Overview on Green Building Label in China. *Renewable Energy* 53: 220–229.

Zheng S.Q., Wu J., Kahn M.E., Deng Y.H. (2012) The nascent market for green real estate in Beijing, *European Economic Review*, Vol. 56, issue 5, pp. 974–984.

7.9.1 Websites

Archdaily (2014) Passive House Bruck/Peter Ruge Architekten (Internet). Available at: https://www.archdaily.com/569638/passive-house-bruck-peter-ruge-architekten-2.

Archdaily (2017) Sino-Italian Ecological and Energy Efficient Building/Mario Cucinella Architects (Internet). Available at: https://www.archdaily.com/880371/sino-italian-ecological-and-energy-efficient-building-mario-cucinella-architects.

Architravel (2013) Sino-Italian Environment & Energy Building (SIEEB) (Internet). Available at: http://www.architravel.com/architravel/building/sieeb/.

CSET website: Centre for Sustainable Energy Technologies. About CSET (Internet). Available at: http://www.nottingham.edu.cn/en/cset/aboutthe-centre/about-cset.aspx.

Floornature Architecture and Surface. Centre for Sustainable Energy Technologies (Internet). Available at: http://www.floornature.com/centre-for-sustainable-energy-technologies-cset-ningbo-china-mca-architects-1681/.

HHD_FUN (2008) Tianjin Eco School (Internet). Available at: http://www.hhdfun.com/200802tianjinecoschool.

Gensler Design website. Shanghai Tower (Internet source) Available at: http://du.gensler.com/vol6/shanghai-tower/#/. (accessed on 12 August 2017).

Hangzhou Low Carbon Science and Technology Museum China (2017) Venue Information: Low Carbon Life, For Our Future (Internet). Available at: http://www.dtkjg.com/pacific_coliseum. (accessed on 22 August 2017).

HHD_FUN (2015) Binhai Xiaowai High School (Internet). Available at: https://www.archdaily.com/773656/binhai-xiaowai-high-school-hhd-fun.

IVL Swedish Environmental Research Institute (2016) Sustainable Economic Growth: Low-Carbon Transition Road Hangzhou Pilot Project (Internet). Available at: http://pocacito.eu/sites/default/files/LowCarb_Hangzhou.pdf. Accessed 11 November 2017.

Kwok N. (2015) HHD_FUN develops vast sustainable high school in Tianjin China (Internet). Available at: https://www.designboom.com/architecture/hhd_funbin-hai-xiaowai-high-school-tianjin-china-09-13-2015/.

Mario Cucinella Architects. Centre for Sustainable Energy Technologies (Internet). Available at: http://www.mcarchitects.it/project/cset-1.

Passive House Bruck website. Available at: http://www.bigee.net/en/buildings/guide/residential/examples/building/27/.

Rodriguez J. (2017) Shanghai Tower: China's Tallest Skyscraper (Internet). Available from: https://www.thebalance.com/shanghai-tower-china-s-tallest-skyscraper-844664.

Schoof J. (2015) Passive house in the subtropics: Residential building near Shanghai (Internet). Available at: https://www.detail-online.com/article/passive-house-in-the-subtropics-residential-building-near-shanghai-16841/.

Tetra Tech. Shanghai (Internet). Available at: http://www.tetratech.com/en/projects/shanghai-tower.

Vyas K. (2016) Binhai Xiaowai High School | HHD_FUN (Internet). Available at: https://www.arch2o.com/binhai-xiaowai-high-school- | -hhd_fun/.

CHAPTER 8

Future Eco-Development in China and Beyond

8.1 Introduction

In recent years, China has experienced a significant shift of focus in its development mode for urbanisation and urban development. Also in these years, most Chinese cities have gone under significant restructuring, some of which have been substantial in modernising the cities. However, the rapid process of development has either already ended or will end soon. Most cities will enter a new stage of gradual development. A similar development trend has already passed to less developed areas and smaller cities and towns, as embedded in the action plans of China's New-type Urbanisation Plan (NUP). Most importantly, however, some of the models and implementation scenarios are either practiced or exported to other countries of the global South, particularly in the regions of South East Asia and the Pacific and in the African context. This also points out the role of China as a successful model of development amongst the developing nations. Nevertheless, Chinese cities are yet to develop further and are still in the process of transitions; whether it would be eco- or green, smart or low-carbon.

If we track the transitions and restructuring of major Chinese cities over the past two decades, it would simply seem incomparable to developed cities of North America and Europe. In a way, we can argue that Chinese cities have benefitted significantly from the ubiquitous trends or concepts of eco- and green development. Why is that so important? And why do we see significant transitions in Chinese cities? The answers have already been highlighted in this book, as we gave examples of innovative projects as experimental cases of development, new Sino-foreign initiatives as platforms of knowledge sharing or knowledge transfer, and new directions of development, indicating policy reforms and transitions, from 'business-as-usual' cases to global cases of sustainable development.

It is important to highlight that China has begun its process of eco-development at a later time compared to developed countries. However, over a very short period of time and particularly over the course of the past two decades, it has steadily gained momentum in terms of policy reforms and policy framework development. During this period, it has also gained attention from the central government and local city officials, which ultimately indicates the importance of the eco-/green agenda at various governmental levels. To avoid the high-carbon paths pursued by developed cities during their early development stages, some Chinese cities start to adopt best urban practices (from various global examples) and develop a multiscalar approach to address sustainability at city, neighbourhood/community and building levels.

Furthermore, it is noted that China has learned from global trends of eco-development, from the very beginning of developing its own initiatives. For example, China's Green Building Evaluation System (GBES) initially used BREEAM and LEED models as their reference examples; China has published its own Passive House Standard, mirroring the best energy practice of buildings constructed in Germany. Furthermore, international best practices have entered the country as China provides vast potential of implementation, particularly in the newly built development areas. Today, LEED is the second mostly used tool for assessing building performance in China. At the city level, given the lack of experience in eco-city development and the relatively high associated initial capital investment, some early eco-city initiatives involve international partners. This was mainly done in order to provide support on city planning, urban design, urban management, advanced green building technologies and financial investment. For example, Sino-Singapore Tianjin Eco-city has been benefitted from the Singaporean experience of developing eco-development projects (see Chap. 5).

As a result of a policy shift towards a more eco-development-style approach in recent years, China is now the leading country in (re)shaping the urban environment. It has the highest number of eco-city projects globally, and is the largest consumer of renewable energies. As of 2016, it also has the largest renewable energy capacity. Meanwhile, energy conservation and environmental protection have become strategic industries that are undergoing accelerated development. Apart from the global trends affecting Chinese practice, China has also developed its own vision on eco-development, i.e., the term 'ecological civilisation' that aims to develop a paradigm achieving harmony between humankind and nature and realising social prosperity in tandem. It is hoped that the current eco-development practices in China might provide useful references to other developing countries that are at a similar stage of development. Nowadays, many developing countries look up to China as a successful leading model of development; and potentially a leading model of green development in the near future.

In this chapter, we first look into some of the past and current trends of eco-development, both globally and China-focused. We follow our previous structure of three spatial levels and provide a detailed analysis of eco-development approaches that exist in the built environment. Afterwards, we provide a comprehensive analysis of 'eco-' from a multiscalar perspective. In this section, we investigate issues of climate change and how it effects urban resilience, and, as such, eco-development as our mainstream. We then discuss more about the context of China and precisely explore the importance of the regional context and climatic conditions, stakeholder constellations as part of project development, and traditions as rightful methods for the future. In the last section, we highlight some of the future directions and give an insight into what directions will be suitable in China, if the concept of eco- is going to continue and strive for further development. The concept will not and must not be outdated, but should rather develop into new directions of sustainable development. The examples that we provide in the earlier chapters will simply serve as a forecast to what the future of 'eco-' will be in the context of China, and how some of these may lead into new models of development for the future and also for the existing built environments.

8.2 Past and Current Trends of Eco-Development at Three Spatial Levels

8.2.1 City Level/Macro

Since 2007, there is a great deal of eco-city development projects in China. Since then, nearly all small- to large-scale cities have a green/eco- agenda. Some of the earlier projects are still under construction and due to completion in around 2020 to 2025. These new eco-city projects are aimed to set new methods of city thinking, creating possibilities for the sustainable transitions of cities. Although very new and mostly experimental, these projects provide us a starting point in the field of eco-city development in China.

At the city level, the factor of 'branding' remains a major matter. In fact, such a branding approach can potentially create an ideology of 'ubiquitous eco-cities', where local environmental/ecological agendas are not necessarily the central concern. The focus should rather be in methods of developing the city environments (if not whole cities), in the most harmonious form with the nature. This ideally revolves around creating or adapting the city environments into sustainable models of lower resource consumption, resource reuse, waste recycling and ecologically sound environments; a mode of development that is either inherited naturally or human-produced (Yigitcanlar and Lee 2014). The central question for any development should be a reflection on 'how development takes place'. Minimising the impact of the built environment on the natural environment is essential in answering the question. The more recent methods of nature-based solutions (currently popular in the European continent) signifies the importance of nature as a priority to urban development. The natural or ecological dimension of eco-city development should not be seen as a commodity (if considered as a branding approach) and should rather be utilised as the backbone of development. In the past, we have seen a great deal of development on preserving and protecting the natural environments. Although less attention is given to nurturing and enhancing the nature. In this process of change, the seemingly irreversible development trends, mainly stemming from the post-industrialisation era, impose a major threat to questions of the nature of eco-cities, and what achievements can be expected from them. The term *eco-city* can then be justified as a more ecologically-balanced city.

What we have seen from our city-level cases in this book are a selection of large-scale eco-city projects in the context of China. Regardless of their actors' involvement and governance structure, these new eco-city development projects can be seen as contemporary eco-city cases of China, or in another words 'local lessons learnt' (Cheshmehzangi 2017). It is important to note their lack of maturity at the current stage, but also their experimental nature that has taken place in the past. In this relatively long and ongoing process, there is an increase in awareness. Although the level of attention has not increased as much, current progress indicates transitions on the ground. This increase of awareness is not yet widespread, but it is visible at the local governmental level. This was not the case about a decade ago. The transition has been slow, for several reasons: (a) huge attention has been given to eco-city branding and not as much on actual changes for planning practices and development strategies; (b) eco-city projects are seen as self-sufficient urban zones that are mostly development at a rapid pace; (c) eco-city projects are implemented on a large scale and mainly located on greenfields or reclaimed lands; and (d) eco-city projects have not yet taken on board a comprehensive understanding and integration of multiple dimensions, such as geographical, cultural, economic and social. The environmental dimension alone cannot create a comprehensive framework for eco-city development.

In most city-level projects, there is often a distance between the eco-development plan and the project outcomes, such as final implementations. While it is likely to see planning becoming more diverse, it is not so simple to diminish the role of planning in the process of eco-development, and particularly at the larger scale of city level. While we can argue that eco-city projects in China are physical (i.e., rather than digital in the case of smart cities), there has been no case of a city-wide eco-development so far. However, a rather smaller scale of eco-city projects, mainly in comparison to the scale of cities in China, is mostly successful. Our evidence shows multiple impacts on: (1) Policies—as they change over the years with more emphasis on ecological factors and environmental agenda; a movement that has firstly pushed and then enabled then local governments to change their mindsets and frameworks; (2) Planning criteria—for instance, from 'business-as-usual' cases to more ecologically-balanced or ecologically-friendly projects, a trend that may have completely changed everything, but has questioned previous planning criteria as the starting point to transition; (3) Design certification popularity—where we can witness a rapid increase of green certification projects and a mandatory green/eco- agenda for city developments; (4) Development mode—which we still find slowly progressing, but clearly in a different direction to decades earlier; and (5) Incentives—where attention is given to support and encourage green strategies and eco development cases. Above all, it is important for eco-cities to include an adaptive approach that encourages innovation; and by that we mean a direction that suggests new eco-directions, new eco-strategic plans and new ecologically-balanced development modes.

In general, we can argue that eco-cities are concerned with both process and the sustainable agenda. Globally, and in China in particular, there is an upsurge of eco-city development that put into practice some of the earlier conceptualisations and visions of the eco-city idea. These practical initiatives are seen as part of an urban sustainability agenda (Joss 2011). Through this, eco-city projects would then revolve around the 'convergence of ubiquitous technologies and city planning megatrends' (Yigitcanlar and Lee 2014, p. 108), which neglects the ideals and situations of the local context. Unlike the more recent smart-city movement, there is a higher chance of approaching eco-city development from the grassroots and a consideration of the social dimension. However, this still offers high potential and we can only see a few smaller-scale cases of success in China. The so-called bottom-up approach would then enable not only new directions of eco-city development, but also leads towards behavioural changes in the general public, development of the social infrastructure, the integration of multiple dimensions in the process of transition, and a consideration of human-centric eco-development. In our conclusions, we highlight some of these further.

8.2.2 Neighbourhood/Community Level: Meso

Although very important, the community level of eco-development in China is relatively weaker than the other two spatial levels. This importance is very much based on its small and manageable scale that can provide pathways to scaling up projects. By this, we simply mean the expansion of projects' ideals or the projects themselves, from a small-scale development zone to a larger scale. Unlike the other two spatial levels, at the community level, the social dimension is more robust and appear to be central to eco- or green development projects on the meso scale. This spatial level can therefore be regarded as the most people-centred level of eco-development.

At this level, the education dimension is significant, and feeds into the increasing awareness of the general public. For instance, the recognition and formalisation of simple—but yet effective—eco-/green factors, such as the increase in recycling, rainwater collection, energy saving, renewable energy usage, public transportation usage, natural cooing and passive methods, and etc. Some of these, although more connected with the micro level, are in fact more effective at a communal setting of community level. To elaborate further, we can just map some of these factors at the meso scale: recycling often covers a larger scale even beyond the community level; or rainwater collection and reuse can be more effective if we consider a larger area of outdoor public places, in between spaces of buildings, and larger bodies of surfaces and storage spaces; or energy saving that is more significant at a larger scale than at the building level alone; or renewable energy usage that often requires a larger area for production and storage, and is less costly when it covers a community rather than just a smaller number of buildings; or public transportation that feeds directly into our spatial patterns, urban forms, and land-use planning; or natural cooling/passive methods that are much effective in cases of district energy systems for cooling (and heating). These are just a few examples to highlight the importance of meso scale solutions for eco strategies.

At the meso scale, we also have a good range of typologies, and we have covered at least four of them in this book. We also found that there is no defined boundary of this spatial level, as it can simply vary from a small residential district to a large-scale development zone. Moreover, it can also vary in terms of functionality, i.e., mixed, or singular use. For instance, for a typical residential compound or district, households are adhered to the community level and cannot necessarily become individual models of eco-development. In such a community setting, the impacts on the ecological infrastructure include the three attributes of building, neighbourhood and location (Yang et al. 2016). Similarly, a typical eco-industrial park is often characterised as a compact area of industries and enterprises having shared resources, operating with synergies of sustainable industrial development, with a low-carbon agenda (Qiu and Huang 2013; Gonela et al. 2015; Xiao et al. 2017). This method of industrial development serves, therefore, as a replacement to sustainable industrial practice (Bansal and McKnight 2009). From its clustered nature of development, the neighbourhood level propose for more implementable strategies of eco-innovation. As such, we can argue in favour of policy instruments available at this spatial level, which improve environmental quality (e.g., very visible for the case of eco-industrial parks), and can ultimately boost technological changes and transitions towards environmentally-friendly development.

One major policy change can be connected to development of benchmarking and rating indicators at the community level, particularly for residential models. In the last decade or so, in China, many new benchmarking criteria are put in place for new development projects across the country. Their lack of presence before indicated 'business-as-usual' for developers, which is now very different in major first- and second-tier cities in China.

The benchmarking at the neighbourhood level, often carried out with rating systems, is an added value of sustainability to any project of the same scale. Simple factors that are highlighted in these rating systems can create tangible and widespread changes. For instance, there is a linkage between household energy consumption and the enhancement of living conditions (Yang et al. 2016), enabling a better reflection on people's lifestyles, and ultimately behavioural changes. It also can create pathways of transitions towards sustainable development, as such, we can refer to the transition of industrial structures, from traditional and high-polluting industries to low-carbon and high-tech zones.

We can see not only policy changes and mechanisms in place, but also changes to planning criteria and design specifications. For instance, we can refer back to the idea of eco-cell development of Sino-Singaporean Tianjin Eco-City, which in fact is a planning model at the smaller scale of neighbourhood level. Also, unlike the current examples of macro-scale projects, the neighbourhood level provides more potential to implement retrofitting projects that include a higher chance of energy efficiency and resource saving (Shahrokni et al. 2014). In this respect, as expressed by Weber and Reardon (2015, p. 114), eco-districts often meet three core goals: '1. To act a basic urban developments that add to the building stock, meeting the need for housing and business space; 2. To include environmental goals going beyond current legal mandates; 3. To test and promote a wide array of eco-innovation: particularly in terms of building and infrastructure engineering and design, as well as new business models and planning processes'. In this respect, we can see substantial transitions of policies and planning criteria. For example, in China, there is now a major agenda on low-carbon transitions of industrial zones towards the development of a green economy. It has also become mandatory for major cities to achieve green development goals and use rating systems in any new development of public buildings. This will soon cover other typologies, and particularly residential as the majority. These directions provide us with a clear understanding on the growing awareness from multiple levels of the government in China, and particularly the local governments that require to interpret and utilise the national-level agendas (Cheshmehzangi et al. 2017a). Finally, we can see many new mechanisms at the local level, focusing on incentive opportunities, new policy frameworks, new local or localised benchmark and rating systems, and multipartnership collaborative practices of eco development at the meso scale.

8.2.3 Building Level/Micro

Energy consumption is largely dependent on the development of socioeconomic and improvement in the quality of life. With the increase in living standards in China, the building energy consumption has been growing dramatically in the past twenty years at an annual growth rate of 10%. People have been demanding more energy consumption to provide a more comfortable indoor environment, e.g., space heating and cooling, hot water, and high-power household appliances.

In 2014, the Ministry of Housing and Urban–Rural Development (MOHURD) developed a long-term roadmap, which set out a time schedule for building energy efficiency improvement during the period 2016 to 2030. In the roadmap, developing net (or nearly) zero-energy buildings (ZEB) become a key target to be achieved by 2030 (Feng et al. 2016). The Passive House Standard, originally developed in Germany in 1990s, is the world's most rigorous energy-based building standard. It requires maximum 15 kWh/m^2 per year for heating or cooling demand in buildings to maintain a comfortable temperature at around 21–22 °C (see Chap. 3). Passive House is widely seen as part of the drive to achieve nearly zero-energy buildings (NZEB) (Wu et al. 2015). In November 2015, China released its own version of the Passive House Standard, the 'Ultra-low Energy Passive Green Building Technology Guide', marking a significant step towards zero-energy buildings (Xu 2017). This new standard is perceived to be an intermediate stage towards achieving net zero energy by 2030, and is more economically feasible for the current economic and cultural context of China (Wu et al. 2015). More discussion on ZEB and NZEB is given later on in this chapter.

In addition, a few key issues should be also addressed in the next step of building energy conservation in China. These issues include increasing thermal comfort in the hot summer and cold winter regions, increasing building code compliance and improving energy efficiency in rural buildings.

China's hot summer and cold winter (HSCW) region is one of the five climate zones in the Chinese building thermal design standard. This region is mainly located in the Yangtze River Basin with a high temperature and humidity in the summer, cold temperature and low solar radiation in the winter. The hottest month usually has an average temperature between 26.5°C and 30.5°C with approximately 80% humidity. The coldest month usually has an average temperature between 0°C and 8.6°C with less than 750 MJ/m² or even 400 MJ/m² solar radiation. The region covers an area of 1,800,000 km² with a population of 550 million people. It is the most populous and economical-developed area of China, producing 48% of the gross domestic product (GDP) of the whole country. This particular climatic region, similar to the other four regions, require specific and localised sets of standards and policies. Therefore, a homogeneous approach to eco-/green building strategies is simply unfeasible in the case of China. For historical reasons, buildings in this region are not well insulated and do not have centralised urban heating systems in winter, according to the Chinese design regulations. These lead to a low indoor thermal comfort in this region. The average indoor temperature in winter is usually below 10°C, and often above 32°C in peak summer (China Society for Urban Study 2016, p. 93). People have increasingly relied on local electric heaters and air conditioners to improve their indoor thermal comfort. Because the buildings are poorly insulated and the heating and cooling equipment has low efficiency, energy consumption has increased substantially in recent years in this region, especially with regard to heating demand.

There is a major debate in recent years about whether urban centralised heating systems should be installed in this region. However, many experts have different opinions. During one interview, Yi Jiang, a member of the Chinese Academy of Sciences in the area of building energy conservation, argued that providing centralised heating systems to the cities in the Yangtze River Basin is both environmentally unacceptable and economically unfeasible. He insisted decentralised or localised heating installations are the future way for increasing indoor thermal comfort in the region. Building energy efficiency technologies, specifically for this region, have been listed as a key area of innovation by the MOHURD in the 12th Five-Year Plan (FYP).

In recent years, China has been forced to achieve higher-level compliance rate of building energy codes to improve the building energy efficiency. Addressing code compliance is required and will be reviewed and approved by the local construction authorities. The MOHRUD has established an annual inspection system to check the compliance rate. Official inspection data show that both design and construction compliance rates have been improved, from 53% and 21%, respectively, in 2005 to 100% and 95.5%, respectively, in 2011. However, these extremely high compliance rates, argued by Bin and Nadel (2012), should be regarded as an indicator of the compliance status of new medium and large building construction projects located exclusively in urban areas, and they do not represent the compliance status of the general population of new buildings in China. The main reason is lack of a clear definition of compliance rates, and a narrow scope of inspection. Nevertheless, the official building energy code compliance rates would seem to indicate a positive, upward trend of improvement in recent years.

Some news reports also show the actual compliance rate can be much lower than initially anticipated. The current practice has been focused on reviewing energy-efficiency proposals at the design stage. However, changing design in the construction stage is often found, for example, increasing the window-to-wall ratio to make the building more attractive—this may accordingly compromise heat loss coefficients and thus violate the building energy code. It is estimated by some experts that the actual compliance rate may be less than 50% (China Property 2017).

Furthermore, the low compliance rates in recent years imply a large share of existing building stock, which is not energy efficient; particularly those buildings completed before the energy efficiency code came into effect for residential buildings in 1995 and for public buildings in 2005, respectively. Retrofitting of the existing building stock in the urban areas is a more daunting task facing the Chinese construction industry in the near future, and was regarded as the next step of policy priority implemented in the end of the 13th Five-Year Plan period.

Rural buildings have accounted for 51% of the total building area in China, and accounted for 39% of total building energy consumption (He et al. 2014). Rural buildings consume more energy than urban buildings to achieve the same level of comfort. This is attributed to several reasons: lack of building codes, lack of professional planning and design, lack of insulation, low-quality materials, and low-quality construction techniques. Energy efficiency in rural buildings is still largely overlooked. Not until 2013 was the first national building energy code for rural buildings—Design Standard for Energy Efficiency of Rural Buildings—issued by the MOHURD. Later, the requirements for energy efficiency of rural building were mentioned in the 12th Five-Year Plan for Building Energy Conservation. However, a comprehensive policy package of incentivising energy conservation in rural buildings is still lacking. Without this, building energy consumption in rural areas, will grow continuously along with the increase of living standards.

8.3 Benefits of 'Eco' from a Multiscalar Perspective

8.3.1 Urban Resilience to Climate Change

Climate change impacts on cities include more heatwaves, extreme rainfall and intense cyclones, harsher fire weather on peri-urban fringes and more severe storm surges associated with rising sea levels. The impacts of such change on the built environment and major infrastructure networks like transport and energy could have immediate and damaging effects on urban communities, the urban environment and a city's productivity (Norman 2016). The role of urban resilience appear more important than ever, particularly that it directly addresses the goals of sustainable development in practice.

The new US administration pulled out of the Paris Climate Agreement in April 2017, which is a major international climate change mechanism. While the decision casts uncertainty over US efforts to rein in emissions, China, the world's largest emitter of greenhouse gases, released its 13th Five-Year Plan to control GHG emission in November 2016. This indicated China's commitment to do its part in combatting global climate change. The key paths are to ramp up renewable energy and curb its use of coal. It just halted the construction of 103 new coal-fired power plants, and its energy agency at the start of the year announced plans to pour more than $360 billion into renewable energy by the end of the decade (National Geography 2017). As part of the Paris Agreement, China aims to peak its carbon emissions and get a fifth of its energy from non-fossil sources by 2030. Some reports suggest it is already ahead of schedule on the former (National Geography 2017).

In order to achieve these targets at local levels, there is a need for further detailed action plans on eco-development at all three spatial levels: city, community and buildings. A broad range of adaptation policies and management strategies has emerged rapidly for managing the risks of climate change. However, their success depends on their uptake by local communities, which to date remains a significant challenge according to some surveys in China. Overcoming barriers to community engagement with adaptation policies, and particularly with follow-up actions, will remain a challenge in some communities until a particularly severe climate event delivers a warning (Cunningham et al. 2016). At the city scale, especially in dense precincts like the central business districts (CBDs), hard surfaces of a city can further increase the impacts of warmer conditions through the 'heat island effect' (Norman 2016). Urban warming carries both health and economic penalties since the urban heat island (UHI) effect reinforces the influence of global warming. Infrastructure is particularly at risk from heating as well as other outcomes of global warming.

Many antidotes to anthropogenic urban heating have been proposed. Technology-based solutions include the use of phase change and photosensitive materials as well as reflective and cool roofs for buildings (e.g., Santamouris 2015). More conventional approaches, being applied widely, consist of increasing the quantum of green space in cities generally, as well as adding green roofs and walls to urban areas. Improving the ratio of ground-level, green open space is a particular challenge given rapid increasing urbanisation and rising densities in countries such as China. Theeuwes et al. (2014) used a meso-scale model to measure the impact of green vegetation and water surfaces on the urban heat island of several Dutch cities. They found that each 10% increase of vegetative cover reduced temperatures by 0.6°C while Coutts and Harris (2013) suggest that a 10% increase in vegetation cover could reduce daytime urban surface temperatures by 1°C. Increasing green coverage is seen as a key strategy to improve urban performance in some Chinese cities. For example, Guiyang is implementing planting programmes in the prioritised areas, including bare mountains, central urban districts, railway corridors and highway corridors.

8.3.2 The Regional Context

As discussed earlier, the context of China is comprised of five very distinctive climatic zones; and this alone can create complexity for eco-development strategies across the country. This means we cannot necessarily foresee a single model or set of policies for eco-development strategies. In contrary, this can partially reduce the ubiquitous character of eco-development projects, depending on the local context and climatic conditions. This can perhaps be seen as a positive attribute for eco-development projects in China. Hence, we see a variety of projects happening in various parts of China that often differ from each other depending on their locations (i.e., being located in cold climatic conditions of North or in hot and humid regions of South China).

An eco-development of any scale has its own barriers and drivers. In the earlier round of eco- projects in China, the lack of sufficient legislation and political guidance is evident. Over time, this has changed to some extent, and policies have become more supportive of eco-development strategies over the years. There is also a potential high cost in any of these eco-developments, especially that some projects are still at their experimental phase. At a smaller scale, the high cost of refurbishment projects does not attract developers or project managers for their practice methods; hence most of the current projects are either renewal projects or completely new (e.g., over previous greenfields, marshlands, brownfields, and etc.). However, our evidence shows that improvements in resource efficiency can have a positive impact on the market value of the projects. Moreover, in general, we see a high demand for green development (such as green certification projects), but less user acceptance in comparison. The economic performance is still questionable for both users and developers, and hence long-term planning is required to provide specific stakeholders with directions and future profitability. Most large-scale eco- development projects are aimed to create welfare and new healthy environments in which people to work and live. However, there is still a lack of progress in reallocating people to new eco-city projects or districts. As a result of this maladministration of population growth, some of the large-scale projects are either deviated into real estate projects for people's investment or have not met their initial target plan. Although it is questionable if the ecological impacts are minimal, we can at least argue in favour of reduction of GHG emissions, better energy use patterns, reduction in material use and resource, and ultimately reduction of ecological footprint through the process of construction and post-occupation.

Nevertheless, some of these above challenges come directly from the lack of regional thinking or regional planning. For instance, for the case of eco-cities, we see a trend of new urban and suburban enclaves (Hodson and Marvin 2010) that are often designed for the wealthy end users. In some cases, the new projects are become the ecological zone of their nearby city or township area. Most importantly, however, it seems worrying to see their lack of integration with their surrounding and nearby development areas. Apart from their physical connections, mainly through new transportation links, there is a lack of regional connections, such as for social development, energy efficiency plans, and environmental factors. The main regional factor that we can see in most cases is the economic factor, which is often related to the linkage with nearby industries. As a result, some of the projects are then built as transit-oriented development (TOD) projects, and often appearing as a business core or residential centre between several nearby industrial zones. At the neighbourhood level, we also see a lack of integration between one community and its nearby blocks or urban environments. Once again, the eco-community project may then appear as an ecological enclave and not necessarily creating a major impact on reduction of environmental impacts and ecological footprints. This depends, of course, on the location of development, and whether or not it is surrounded by high-polluting neighbourhoods. For the cases of micro-level, eco- and green buildings often appear as new models of development (if not landmarks) in their vicinity. This can, in the long run, have a positive effect on trends of development in a particular urban area. However, such an effect is more likely if the same developer or a same group of stakeholders get involved in any nearby development projects.

In summary, we would like to highlight the significance of regional context in eco-development projects. The meaning of 'region' or 'regional' is dependent on the context of development, and how it is interpreted in its process. This can mean climatic conditions, or creating a network of people mobility, and even a possibility to encompass strategies for material use reductions, carbon reductions, and integrated energy systems on a large scale. By doing so, the strain from urban competitiveness may also ease, which can ultimately allow for a network of ecological zones and eco-development projects, rather than just development of a singular ecological enclave amongst a region of other development projects. The questions here are that: are we really living in the *Anthropocene* epoch? And, if so, are our relatively small experimental projects creating enough impact to reverse some of our past and current unsustainable trends? And what future directions are ahead if eco-development is ever going to last longer and progress? In the next section, we aim to address these questions by simply providing some future directions, few of which we find as necessities for the future of China's eco-development.

8.3.3 *Stakeholder Constellations*

Balancing the profitability across various stakeholder is always a major challenge of any development. It is even more complicated in the process of eco-development. However, there are often signs of motivation and vision of those developing and living in eco-communities (Lovell 2012, p. 3) that include: 'the desire to distinguish themselves from others by actually engaging in environmentally and socially sustainable lifestyles and practices; staying ahead of anticipated changes to housing policy and regulations; and the opportunity to exploit new business opportunities. Motivations vary according to the type of institution involved (community-based, social housing, private housing developer, government, etc.'. Through our case study research studies, the highest level of desire is witnessed at the local governmental level. The main drivers for this desire come from the main factor of urban competiveness, as well as feeding into the national-level agenda and frameworks, and increasing the growth of GDP in comparison to any nearby rival cities, districts or towns.

In China, and as show in our case study sections, the eco-development projects include multiple actors and stakeholder constellations. For instance, at the city level, we have recorded a range of eco-city projects from local development, to national flagship projects, and to international or joint cooperation; hence our case study categorisation is focused on this situation of actors' involvement. Similarly, at the neighbourhood level, many projects (e.g., eco-industrial parks, eco-villages, and eco-community projects) include international partners or are jointly done with various local actors. At the building level, the stakeholder constellations are less complicated and may include multiple stakeholders (such as international actors) only depending on the status of the building or the importance of the project at a higher level. However, in contrast to the other spatial levels, building projects are not regarded as Sino-foreign collaboration projects.

In general, and regardless of spatial levels, the involved stakeholders have their own set of interests that may often be in conflict with one another. In most cases, we have seen more overlaps between them than actual conflicts. In this section, we consider some of these stakeholder constellations, namely governmental authorities or governance, developers and project managers, planners and designers, and communities and the general public.

8.3.3.1 Governance
The role of authorities and governance is very much in evidence in China's eco-development projects. This relies heavily on the development of policies, frameworks and best practices that are often developed as experimental projects. In this process, there is a major need for the introduction of innovative forms of governance, such as the existing model of multifaceted governance. The relationship of governance to both market and society is

dependent, then, on future directions, potential reforms and policy developments. To achieve a genuine eco-development, regardless of its spatial level, a visionary leadership is highly required and recommended. The ideals of *eco-* would then lead towards the development of new ecosystems. In China, we anticipate the emergence of a more matured-up local governance set up in the coming year. The maturity is already visible in most cases where transitions are already progressing and where national-level policies are interpreted, contextualised and adapted to the local demands. This adaptability characteristic of the local governments promotes their position as key actors of change and transition.

8.3.3.2 Developers and Project Managers

Very much similar to any other country, developers play a major role in the delivery of eco-development projects in China. These can include local developers, state-owned developers and international developers. Their role is important, since they are not necessarily the end users, but often just developers of the projects. In our studies, we found out that multi-partnership projects are often more successful in their delivery, since such a mechanism enables the developers to strengthen their strategies, save costs and increase their revenues (EURObiz 2013; Cheshmehzangi et al. 2018). In smaller-scale projects at the meso and micro levels, this may be more difficult to achieve as multiple partnership is often narrowed down to a team of project contractors and subcontractors. More importantly, related regulations and policies in regard to green and eco- development should be in place to allow for a greener development. Without these, developers would select the 'business-as-usual' scenarios and often go with the most profitable options. To avoid this, we stress the requirement for comprehensive and directive planning and design regulations, monitory mechanisms and incentive opportunities (Cheshmehzangi et al. 2017a). The effect should potentially encourage innovative thinking for green and eco- strategies amongst engaged industry stakeholders and towards the promotion of sustainable development.

8.3.3.3 Planners and Designers

The role of planners and designers is perhaps the most difficult in practice. Their positioning between the developers and end users makes their role even more challenging. However, they are also in the professional position to advocate the authorities and developers, and suggest alternative design and planning solutions to cases of 'business-as-usual'. It is perceptible that over the years, the Chinese practices and China-based foreign practices have grown maturely by undertaking some of the concepts from integrated design thinking, urban ecology, landscape planning, green assessment techniques, modelling simulations, and so on. Their global network and their professional insight in new development opportunities, puts planners and designers in a stronger position to define and refine the pathways to sustainable development. Globally, we can see substantial change in the role of planners and designers as part of the sustainable development agenda, highlighting the fact that theirs is a flexible role in the process of urban change and transition. In China, we anticipate to see a more effective role of planners and designers in future green and eco-development projects.

8.3.3.4 Communities and the General Public
This group of stakeholders is generally known as end users. Yet what we see globally is a change to this discarded thought and towards an ideal that communities are more than just the end users. The importance of grassroots and human-centric development, has been one of the central arguments of the national government in China. Also we can see traces of such change in China's national New-type Urbanisation Plan (NUP), since its inception in March 2014 (Cheshmehzangi 2014). Therefore, there is still room for policy development and enhancement of community-oriented strategic plans. As described by Shi et al. (2016), proper policies can motivate the participation of the local community in the sustainable neighbourhood development. This will then lead to a strengthening of the the institutions, community engagements and public participatory scenarios. It is also argued that with such policies in place, we can encourage the community residents and representatives to take a more active role in recognising their responsibilities for sustainable development and enable them to contribute as active users (Zhang 2011) and not just end users. Hence, it is always important that in China we consider the role of communities and community representatives central to achieving eco- and green goals of sustainable development.

8.3.4 Traditions

In ancient China, people constructed their cities and building based on the principle of 'harmony between man and nature'. It refers to build a harmonious relationship within human, architecture and nature, which exactly meets the value that the green development aims to deliver today. The wisdom of ancient Chinese architecture is to extract essence from nature, imitate the ecological environment, reduce building energy consumption, and truly achieve the purpose of energy conservation. In addition, the ancient architecture has a simple layout, with particular attention to the surrounding natural environmental coordination. Thus, at present, the knowledge of wisdom from ancient Chinese still helps to build modern and environmental-friendly architectures. For example, Jiangnan Garden—a traditional Chinese construction in the Southern Yangtze River region—brings the mountain, water and forest of nature into the courtyard. Meanwhile, people feel the beauty of simple, fresh and nature (Chen and Wang 2012). It is often much appreciated amongst the general public. Generally, traditional Chinese architecture advocates simplicity. The idea plays a guiding role in the construction of modern buildings, social life and human value.

Evaluating the cases we discussed in the earlier chapters indicates the current eco-developments in China are largely influenced by modern urban planning theories such as new urbanism and compact city, featured with mixed-use neighbourhood, effective public transport, etc. They originated from the western context, reflecting the advanced planning and design knowledge today. However on the other hand, what can we gain when we look back to the Chinese traditional architecture practice, which is largely ignored in those case studies? The Li-fang system has been an ancient urban form in China for thousands of years since the Han dynasty (202 BC). A whole city was divided into a number of Li-fang units and each of them acted as a specific functional zone, such as residential, commercial, religious and political. For example, Chang'an city (the present-day Xi'an city) had 108 Li-fang units (Zhang et al. 2014). The city was constructed as a square, with three gates placed on each side of the boundary wall connected to main streets horizontally and vertically (Zhang et al. 2014). Another example is Suzhou city, which was originally built in 514 BC and experienced a major redevelopment in 248 BC. The main feature was to optimise the waterway network to facilitate transport and protect the city from floods. Since then Suzhou had been away from major flooding events for hundreds of years.

We notice that most residential neighbourhoods in China are enclaved 'small residential districts' (SRD), which is the housing form for the majority of Chinese urban residents. It is a planned neighbourhood where housing is integrated with communal facilities such as kindergartens, clinics, restaurants, convenience shops and communication infrastructure—all under the control of a professional property management company. Some so-called 'super-blocks' may constitute thousands of buildings. For example, Tiantongyuan in Beijing covers eight square kilometres, has over six million square metres of building area and was home to 400,000 people by 2011. Non-residents have to drive around Tiantongyuan, rather than through it. A resident living in the compound may have to walk 20 minutes on its boundary (China Daily 2016). This type of neighbourhood, as argued by Deng et al. (2014), can incur resources waste and urban traffic congestion. Ancient Chinese urban planning and design may shed some light on this issue.

At the building level, we also notice that traditional architecture forms are largely neglected. Here we present two traditional Chinese architecture examples to show what we can gain by looking back to the traditions: China Patio House and Cave Dwelling.

8.3.4.1 *Chinese Patio House*
The conventional Chinese house roofs have different slopes, resulting in various forms of courtyard-style residential typology. That means the houses of four sides (or sometimes three sides) are being connected together and surround a small courtyard in the central. Around the core patio, we find a hall, bedroom, kitchen and other utility rooms. The form of residence is the so-called 'Patio House'. Outside the patio, there are corridors, railings and decorative openwork windows, causing the change of light surface. Although the houses are enclosed together, this allows sunshine and rain to reach the ground. The demand actually reflects the close relationship between human and nature. For most traditional houses, the patio plays an important role in the daily ventilation and lighting. The patio allows air to flow through and, combined with the open hall and corridor, it forms a ventilation system. To be specific, the Sun illuminates the house during the daytime, leading to the patio temperature rising fast. Then the air in the patio keeps rising and the air pressure is falling. While due to the rooms being covered by roof and floor slab, the air temperature is relatively low and pressure is large inside the rooms. Thus, hot air rises in the patio and cold air from rooms constantly replenish the patio area. This effect increases the air convection between patio and internal rooms. At the time, the patio area, operating like chimneys, has the function of

pulling up wind and forming hot pressing ventilation (Chen 2014). At night, the air cools down quickly over the patio; thus, the patio is regarded as cold source and the air pressure is relatively low. However, the indoor cooling is relative slowly, resulting in the higher room temperature and lower air pressure. Therefore, the cool wind flows from patio to each room and the water inside the patio makes wind cooler.

In addition, the surroundings of patio are generally transparent and rooms on both sides are decorated with carved windows, which is effective for ventilation and lighting. At the aspect of rainwater treatment, the roofs of 'Patio House' are normally in the inner slope. Rainwater flows along the roof to the patio, through the rain pipe on the roof comes down to the ground, and then be drained out of the house via the gutters around patio. To sum up, the small patio can prevent exposure to the Sun and keep the house cool in summer. In rainy days, residents will not feel wet due to ventilation. Hence, 'Patio House' shows the full application of natural conditions to create a healthy living environment (Chen 2014). More importantly, such a typology of housing represents one of the best typologies of person-centred development.

8.3.4.2 Cave Dwelling

Cave dwelling is an old way of living, which means to excavate the transverse cave in the loess cliff area as the living rooms. Different from normal traditional Chinese architecture examples, it uses local raw soil as the building construction material instead of wood. Thus, the cost for material transportation can be saved, as well as the construction cost and the use of wood. Since the envelope of caves is loess, it has minimum heat loss and is well insulated. Besides, the shape of a cave dwelling shows arched and its opening generally facing sun; hence, the solar radiation can be fully used for the indoor space (Chen 2014). Thus, it is warm in winter and cool in summer, and is a natural energy-saving building. In addition, the thick soil walls of caves also have a sound insulation effect, which effectively avoids neighbourhood interference. The cave dwelling is simple to construct and durable, it saves arable land, protects vegetation and it does not destroy the ecological environment. Therefore, even at present, cave dwellings are still widely used in the loess area.

In summary, it is worth to look back the traditions and this may give us a different angle of viewing eco-development. Modern architecture, largely transplanted from the West, will be beneficial from such retrospect.

The examples discussed above simply highlight the importance of vernacular thinking and the context in which we should identify and work with certain historic, cultural and environmental characteristics.

8.3.5 Breaking Misperceptions

Technology will not provide all the answers to problems of hyper-urbanisation, but it will bring significant improvements in energy and resources use and will encourage accurate measurement of our progress (or regress) in the quest for sustainability. Clear directions of movement will lend themselves to more clear-cut responses too. Nevertheless, there are still misperceptions regarding eco-development among various stakeholders. It is more a case of obtaining behaviour change by breaking misperceptions.

Developers do not recognise the market returns for investment on the green building development. The most significant factor influencing their decision making is still the incremented initial construction fees. For example, in order to attain green building certification, they need to spend more on energy, water, renewables and other technologies. Currently, this increment is higher than the government subsidies. Furthermore, even they are aware of the lower running cost and other benefits in the long term, developers cannot benefit from the investment on green development, since developers are usually not building managers and therefore are not beneficiaries of less maintenance, higher rent and higher productivity.

Many property buyers or renters also have misperceptions about green buildings. The reasons leading to the phenomenon can be divided into two aspects. The willingness to pay higher for green buildings is still low according to a survey conducted in Nanjing that was discussed earlier on in Chap. 4. In addition, buyers do not have sufficient knowledge about green development as it is still a new concept in China. Many of them think green development refers to large lawns and beautiful landscape. The financial return in the long term is not yet part of their buying decision.

Fostering a green development market is the key to mobilise relevant stakeholders. On the one hand, the government should fully mobilise developers to play the guiding role in the development by increasing both regulatory requirements and subsidies. On the other hand, promoting knowledge on green development, especially the financial feasibility, will have significant impact on buyers' behaviour.

8.4 Prognoses for the Future

8.4.1 Emerging Trends

Today, most countries, even in the developed world, are still fixated with economic growth as the fundamental criterion for measuring success in society. Thus, when there are conflicts between development projects and environmental goals or social fairness, economic interests are usually favoured. However, at a different level, there is considerable hope that we will move on gradually to a more sustainable pathway. Perhaps a key to our 'redemption' is the advancement of awareness, policy and technology for moving our cities, communities and buildings forward in both developed and emerging nations.

Recognition of difficulties with and pressures from nature's limits comes from all sectors of society. Corporate awareness of the concept of sustainability is growing and as many as 21 stock exchanges across the world could introduce sustainability reporting standards in the coming months. They would join the 17 exchanges that currently recommend listed companies report on environmental, social and governance (ESG) issues. The exchanges go a step further by providing model guidance to participating companies (Khalamayzer 2016). In addition, many managerial elements of government have engaged with goals that pursue sustainability and non-government organisations, academia and interested parties in civil society are similarly concerned about environmental responsibility and social equity. Intelligent and innovative solutions to our challenges from this host of stakeholders are legion (Deng et al. 2017).

Sustainable urban land development and smart growth policies can have significant effects on reducing harmful emissions and the demand for energy. Empirical studies from built environment-travel literature indicate that sustainable urban form—compact, high-density and mixed land use—can reduce per capita vehicle travel, distance, time budget (Vance and Hedel 2007), and results in lower transport-related energy demand and greenhouse gas emissions (GHGs) (Li 2011). For instance, 'residents living a large city (population between 5 and 10 million) in comparison to a small city (population < 1 million) makes 16.4% more expenditure on private and 26% more expenditure on public conveyance (Ahmad and de Oliveira 2016, p. 110). A 10 percent increase in density is associated with 1–1.2 percent decrease in the amount of transport and 0.5–1 percent reduce in vehicle miles of travel (VMT), keeping other variables constant

(Ahmad and de Oliveira 2016). Hence, urban form along with associated negative incentive programs, such as driving restrictions (e.g., 'no driving day'), high parking fees and high taxes on cars, and congestion charges (e.g., London) have been traditionally considered to reduce the use of private transport. Additionally, provision of monthly pre-paid public transport passes (e.g., Seoul and Tokyo), and improved public transportation and mobility (e.g., bus rapid transit, mass rapid transit, bicycle lanes) have been under implementation to drive cities follow a sustainable path.

Research on the relation between human health and well-being and urban planning suggests that urban environments that lack public spaces can encourage sedentary living habits, while the provision of open spaces can facilitate social interaction and provide opportunities for physical activities (Mytton et al. 2012). The likelihood of being physically active may be up to three times in residential environments that have access to green spaces, and the likelihood of obesity may be up to 40% less (Coombes et al. 2010). Natural landscapes are thus vital to urban liveability, and good urban design outcomes always strive for the 'right' balance between urban development and the preservation of natural environments.

Resilient cities address the capacities of built environment and people to adapt to natural and human-induced shocks/threats, such as earthquakes and climate change (Desouza and Flanery 2013). Cities become more resilient if they better address stresses and mitigate risks. A transition from urban sprawl to eco-city or carbon-neutral city will help to reduce automobile emissions, improve air quality, reduce energy demand, and create a resilient urban environment for people (Joss 2011; Roseland 1997). Key features of an eco-city include: planned walkable communities, green transport, adaptive reuse of old buildings, eco-friendly new buildings (e.g., passive solar design, environment-friendly building materials), green infrastructure (e.g., greenways, gardens, green roofs, waterways and community farms), energy efficiency and renewable sources of energy, and effective waste and water management (reuse and recycle).

In recent years, collective civic structures and innovative models of shared economy have been providing an unsurpassed platform for more socially inclusive and environmentally efficient urban communities in cities. Sharing enhances the potential resource, environmental and energy efficiency of cities, while less waste is created as everyday consumer goods and products are shared among a group (Belk 2014). Sharing could also provide an opportunity for a community to enjoy things at an economical cost and save and make money (Agyeman et al. 2013). Globally, 'urban sharing schemes' have multiplied into different dimensions: bike-sharing schemes in London (the UK), Paris (France), Rio de Janeiro (Brazil), Hangzhou (China), Guadalajara (Mexico), Buenos Aires (Argentina), and Providencia (Chile); car-pooling schemes in Lisbon (Portugal), Stuttgart (Germany) and Melbourne (Australia); or car-sharing services such as Uber, Didi, Spotify, Taxify, Cabiday, and Zipcar; sharing apartments and spaces (e.g., AirBnB, Couchsurfing); sharing and swapping consumer goods (e.g., Facebook Market Place, E-bay, Gumtree, Freecycle); sharing privately owned public spaces, such as Zucotti Park in New York City, Dewey Square in Boston, Canary Wharf in London; and examples of sharing power sources, water and sanitation in informal community settlements in cities such as Mumbai (India), Cali de Santiago (Colombia). Moreover, we can argue that sharing will significantly assist governments and policymakers in achieving standards in eco-, inclusive, smart, sustainable, and resilient cities.

Cities will compete globally to make their urban areas attractive both to live and to invest in, and face the challenging task of balancing between competitiveness, environment and quality of life. Making cities sustainable and liveable is the only way to counter the negative effects of the global megatrends, such as globalisation, urbanisation, demographic change and climate change, which often create environmental, social and economic problems for cities. Urbanisation per se can be a welcome phenomenon of change and transformation if the visions of 'sustainable', 'integrated' and 'resilient' concepts and the relationship between people, built environment, and ecosystems are articulated in defining an urban development trajectory.

8.4.2 Burgeoning Building Technologies

New technologies can also bring a host of new processes and vast increases in efficiency and productivity in cities, communities and buildings. Examples include the integrated network of smart products—the Internet of Things (IoT); smart cities and 'Big Data'; advanced robotics and artificial intelligence; accessible renewable energy harvesting technologies and an energy internet (Rifkin 2014, p. 72). Open online education is gaining adherents, there is the decentralisation (and democratisation) of finance and governance, through Blockchain applications, including its applicability to streamlining micro-power grids operating at precinct and neighbourhood levels. The decentralisation of manufacturing through 3D printing is gaining traction too. It uses only 10% of the material involved in traditional manufacturing as well as eliminating the need for lengthy supply chains.

Many green building projects emphasise on technological innovation as the main means of achieving green building vision that leads to adopt more active measures. Of these, a large majority focuses on energy technologies, including renewable energies. However, the adoption of new technologies in the construction sector has been still slow. A recent report by the World Economic Forum (2016, p. 5) states:

> *The Construction sector has been slower to adopt and adapt to new technologies than other global sectors. While innovation has occurred to some extent on the enterprise or company level, overall productivity in the sector has remained nearly flat for the last 50 years.*

This report further identified methods such as information modelling approaches (at various scales of building or city) and prefabricated building elements as recent individual technologies that have the highest impact on future construction practice (World Economic Forum 2016). Other technologies include the 3D printing of components, advanced building materials, and real-time mobile collaboration. Some technologies, while still in their infancy, have been pioneered in China. For example, China has used 3D printing in conjunction with innovative building assembly innovations involving penalisation and modular components to achieve remarkable efficiency levels in the construction industry. A five-storey apartment building was made in Suzhou entirely with a giant long 3D printer, making it the tallest 3D-printed building in the world.

Emerging technologies such as Building Information Modelling (BIM), City Information Modelling (CIM), prefabricated building and 3D printing are attractive partially because they are promoted by the Chinese government. For instance, in Shanghai, BIM is compulsorily required to be used in building design and construction from 2018. A minimum of building prefabrication is also a mandate for new buildings in Shanghai and Zhejiang Province. The strong technology focus in some cases can be explained by the mainstreaming of and the prevalence of mainly technocratic approaches reflecting current regulatory trends.

Technology will not provide all the answers to problems of built environment being as a major energy and resources consumer, but it will bring significant improvements in energy and resources use. It is important for us to identify and utilise the advancing technologies in sectors of building construction, architecture and city modelling, and urban development. We explore two short cases here that have adopted new technologies representing the future technological trend in the industry in China.

8.4.2.1 Example 1: 3D Printing Buildings in China

In August of 2014, WinSun, based in Shanghai, completed 10 houses under 24 hours in Shanghai using a 3D printer (News Atlas 2014). A year later, in 2015, this company has expanded the capabilities of 3D printing, has built a five-storey apartment block, which is the world's tallest 3D printed building, and a 1100 square metre villa with internal and external decoration in Suzhou Industrial Park in Jiangsu province (3ders.org website 2015). The two buildings represent new frontiers for 3D printed construction, partially demonstrating its potential for creating more traditional building typologies and therefore its suitability for use by mainstream developers.

The buildings were created using 6.6 metres high, 10 metres wide and 40 metres long 3D printer which builds up layers of an 'ink' made from a mixture of glass fibre, steel, cement, hardening agents and recycled construction waste. With this technology, the company is able to print out large sections of a building at WinSun's facility, which are then assembled together on site, complete with steel reinforcements and insulation in order to comply with official building standards (Winsun website n.d.). The raw materials as printing 'ink' are mainly construction waste, industrial waste and mine tailings, and the other materials are principally cement and steel, as well as special additives. Through the 3D printing technology, developers can shorten the time required to build a house, save construction costs and materials, and reduce construction wastes. In an interview with China Daily (2015), the president of Winsun estimated 30–60% building materials and 50–80% labour are saved by using 3D printing technology.

8.4.2.2 Example 2: Information Modelling Techniques in Green Building Design and Operation in China

A modelling approach, such as Building Information Modelling (BIM) is a collaborative way for multidisciplinary information storing, sharing, exchanging and managing throughout the entire building project lifecycle, including planning, design, construction, operation, maintenance and demolition phases. This method has been in practice for many years and in various methods and already under practice as the city-level modelling method as well (e.g. City Information Modelling). The Chinese government has announced BIM-relevant policies and standards to boost the development of BIM during the years of 2011 to 2015. The strategy of BIM application in the AEC sector also requires that by 2016, government-invested projects over 20,000 square metres and 'green' building in the provincial level should adopt BIM in both design and construction stages, and by 2020, the industry guidelines for BIM application and government policy systems should be well established (Cao et al. 2017).

BIM not only brings 3D revolution for traditional 2D drawings of AEC industry, but also emphasises the significance of multidisciplinary collaboration for projects. BIM can be integrated with other building energy simulation systems, such as Energy Plus, DOE-2 and Equest. It is therefore, a representation of an integrated model that utilises a multifaceted approach to building construction, or urban development, including the optimisation of low energy design in the planning and design phases, the efficient reduction of energy waste, monitoring and control of building energy use, and etc. (Habibi 2017).

Bund SOHO is located at the interchange of Second East Zhongshan Road and Xinkaihe Road, Huangpu district. The project was completed in 2015 and has a footprint of 190,000 square metres. It is composed of four office high-rise buildings with heights of between 60 and 135 metres. Bund SOHO is a multiple-use landmark containing urban development, comprising offices, commerce and entertainment, enjoys superior location and splendid views of the Huangpu River. It has a gross floor area of 189,449 m^2 and underground parking and space for plant rooms of 776,909 m^2.

BIM implementation plan of Bund SOHO project was officially launched in June 2012. It is the first project in China which uses BIM technology throughout the whole construction processes. In order to ensure the smooth implementation of this project, the developer, together with the design consulting company and other stakeholders, developed a BIM implementation organisational structure specifically for this project. The participants included design consultants, costs control, project management and IT consultants. Other stakeholders, such as modelling and simulation practitioners, contractors and subcontractors, were also involved. At the end of 2012, under the guidance of the developer's BIM team, the BIM test was carried out on the contents of the basement civil construction to ensure a full BIM application can be achieved for the project in 2013 (SOHO BIM project report 2013).

At the conceptual and early design stage, the weather analysis tool EcoTect was used to estimate and ensure the weather characteristic of Shanghai, to optimise the building orientation and reduce the energy load. The main building orientation of this project is facing south-east, the best building direction of Shanghai. Natural ventilation is utilised in summer and external shading is provided to avoid excess solar radiation; sunlight is able to enter in winter, but the wind intake is optimised to reduce heat loss. Energy consumption, water consumption and carbon emissions for the entire building are analysed and optimised based on the conceptual design model. Green Building Studio estimates are made through Revit Subscription by setting geographical information and various energy options in Revit. The energy-saving rate of this project is greater than 50% (ibid.). Therefore, we can argue in favour of such modelling techniques that enable us to utilise data for more collaborative uses and integrated practices.

8.4.3 Zero Energy Building and Life Cycle Consideration

Developing net (or nearly) zero-energy buildings (ZEB) become a key target to be achieved by 2030 in a roadmap proposed by the MOHURD (Feng et al. 2016). Currently, there is no strict definition of a ZNB or a nearly zero-energy building (NZEB). Nevertheless, a ZEB or NZEB is an energy-efficient construction, which features the following factors:

- Minimises energy demand through passive design, e.g., insulation, natural lighting, ventilation, airtightness, efficient heating and cooling, and etc.;
- A ZEB requires energy that is fully supplied by either onsite or nearby renewable sources, such as PV panels and ground source heat pumps;
- A NZEB requires a small amount of energy from the grid.

Many countries have developed plans to promote ZEB or NZEB. For example, the European Union (EU) has a major initiative, labelled the Energy Performance of Buildings Directive (EPBD), which requires 'nearly-zero-energy buildings' (NZEBs) target for all EU member countries' new construction by 2019. In the UK, all non-residential buildings must achieve zero-carbon targets by 2016.

Apart from declaring a direction for future building development in the roadmap, other details are not publicly available at present. Xu Wei, the head of building energy efficiency in the China Building Science Institute, proposed the following targets to be achieved by 2030: 30% of new buildings to meet NZEB requirements; 30% of the existing building to achieve NZEB requirements; and renewable technologies to supply 30% of energy in new buildings (Xu 2017).

ZEBs is still a new concept in China, policy instruments, such as certification, information tools, demonstration projects, education and training, and R&D are lacking. By the end of 2015, fewer than ten demonstration projects have been reported (Feng et al. 2016). Most of these projects are not certified as ZEBs by a certification system and mainly used to test their design and building components. The first passive construction was built in Shanghai in 2010 and certified by the German Passive House Institute. However, as the ultra-low passive building is listed as a key technology to promote building energy performance in China, it is expected that we will see more passive buildings developed and tested in the next few years. These practices represent a beginning of ZEBs' development in China and send signals to the construction market of the goal of moving to ZEBs in the long term (Wu et al. 2015).

A more rigorous definition than ZEBs is known as lifecycle energy-neutral building, which requires the building itself to generate more energy than the amount required to maintain its operation. The extra energy generated is used to offset the embodied energy used for raw materials extraction, manufacturing of building materials and elements, transportation and construction process. This often refers to the pre-occupation phase in a building's lifecycle.

To achieve a lifecycle energy neutral building, a lifecycle assessment (LCA) must be conducted to quantify and compare material & energy flows and the related emissions of the built environment at different spatial scales and in different contexts. Building LCA analysis is more complex than for many other industrial products because it requires the aggregate effects of a host of lifecycles of their constituent materials, components, assemblies and systems. This method is mainly used for building materials accounting and is emerging at the level of whole buildings. Several recent computer programmes incorporate LCA methods into tools for design and analysis of buildings such as BEES, Athena and Envest. However, due to data limitations, the large range of construction techniques, material and system choices in buildings, none of these tools are currently capable of modelling an entire building, or computing environmental impacts for all phases or processes (Cole 2010, p. 274).

In China, LCA has been emerging as a new concept. Over the last couple of years, it has been moving from research arena to practical implementation. For example, in July 2016, the State Council requires the construction of a green manufacturing technology system based on lifecycle assessment in the 13th Five-year National Science and Technology Innovation Plan; and in December 2016, the State Council released another policy guidance requiring to construct a comprehensive LCA-based assessment system to label products. However, current LCA policies and practice are focusing on industrial products. There is, at present, very little application of LCA in the construction practice in China. The main difficulties include the lack of a database of energy intensities for various building materials and elements, and a lack of reliable software programs for whole building simulation. Some organisations are trying to bridge these gaps. For example, Sichuan University is establishing a database called CLCD (China Life Cycle Database) and setting an online platform, eFootprint, to promote LCA in China.

8.4.4 Nature-Based Solutions

In the past few years, the new concept of 'nature-based solutions (NBS)' have been given significant attention in the European continent. This was initiated by the European Commission in 2013, and, subsequently, it has gained popularity in research over the years. The concept is given a defined place 'within the spectrum of ecosystem-based approaches' (Faivre et al. 2017, p. 509), enabling researchers and practitioners to find innovative solutions with nature to address challenges of sustainability dimensions. A method that can provide us with a holistic approach towards the enhancement of our ecosystems, biodiversity and livelihoods of our environments. It also fits into parts of green economy, providing a range of sustainability objectives, practical implementations, and nature-based strategies.

In their recent studies on 'NBS to address global societal challenges', Cohen-Shacham et al. (2016) suggested five categories of NBS approaches: Eco-system restoration approaches (such as, Ecological restoration; Ecological engineering; Forest landscape restoration); Issue-specific ecosystem-related approaches (such as, Ecosystem-based adaptation;

Ecosystem-based mitigation; Climate adaptation services; Ecosystem-based disaster risk reduction); Infrastructure-related approaches (such as, Natural infrastructure; Green infrastructure); Ecosystem-based management approaches (such as, Integrated coastal zone management; Integrated water resources management); and Ecosystem protection approaches (such as Area-based conservation approaches including protected area management). All these categories of NBS approaches fit directly to the ideals of eco-development, particularly at a larger scale of city/district level, and neighbourhood planning. More importantly, NBS approaches are identified as nature-driven methods to rehabilitate and restore the land as a new method of land management and towards enhancing ecosystem services (Keesstra et al. 2018). This method is based on a system thinking approach, working with innovative solutions derived directly from nature. Therefore, NBS approaches are, in fact, the dynamics of the system (ibid.). Furthermore, NBS is defined as methods of integrated thinking at multiple levels, and often includes a multi-stakeholder dialogue (Nikolaidis et al. 2017) between science, policy, business and society.

In the studies of NBS and the built environment, researchers have so far looked into many cases that ultimately help to improve the urban environments, public health, well-being of the society, economic growth, and the ecosystem services. To name a few, we can highlight recent published studies on: urban natural environments as NBS and their role to improve public health (van den Bosch and Ode Sang 2017), NBS for urban water management and values of green infrastructure (Wild et al. 2017), NBS for well-being in the urban environments (Vujcic et al. 2017), methods of promoting human resilience in cities and against the heat (Panno et al. 2017), and many more. The common factor between all these studies is that NBS approaches are used as tools to achieve sustainable development. Likewise, the concept of eco- or green- should merely be a comprehensive tool to achieve sustainability in the built environment of multiple scales and multiple dimensions.

Although the term eco-development mainly highlights the ecological/environmental dimension of sustainability, it comprises the impacts that environment can have on the other dimensions and vice versa. Unlike eco-development studies, most NBS research is focused on scientific and environmental research and not much on planning. This is anticipated to change as the flow of NBS research meets the needs of urban development, planning strategies, and even eco-development. Moreover, much of existing research on NBS is focused on the European context; but with the growing EU–China collaborations in recent years, the topic is also receiving attention in China. This occurs particularly in the topics of public health, resilience, and climate change which are of national-level interest in the context of China. Therefore, the role of NBS in anticipated to become more valued and visible in China, specifically in the more matured areas of integrated planning from various perspectives (Saleh and Weinstein 2016; Cheshmehzangi and Butters 2017), urban landscape assessments (Fan et al. 2017), marine studies (Kelly et al. 2012), and flood protection studies (van der Nat et al. 2016); all of which are currently embedded in sustainable development strategies of China. The future directions, and as it can be similar for eco-development strategies, can be made based on a combined method of nature-based solutions and technology advancement. The two, if complementary to one another, can provide us with innovative directions towards eco development.

8.4.5 Integration: Concepts of Eco-Innovation and Eco-Fusion

Lastly, what we argue about is the importance of integration in eco-development strategies and planning. This can only occur by consideration of the multi-scalar and multi-dimensional nature of urban sustainability that provides us with models of best practice. Recent studies include various factors and recognition of urban sustainability in practice. This approach can be discussed as part of the Integrated Design Approach (IDA) or purely the integrated thinking in design development of the built environment (Cheshmehzangi et al. 2017b). Based on our case study research in China, we found integration lacking in many aspects. There is still lack of integration of eco- strategies in planning practices, and some of the eco-city projects have evolved into a *greener* real estate projects over the years. It is still anticipated that some of the new planning practices should integrate more of ecological indicators and environmental agenda. However, in some cases we see similar and wide street layouts, high-rise and gated housing compounds, a car-friendly living environments, and strong presence of zoning. All of these issues should have diminished in the process of eco development at large scale. On the other hand, in some cases like Chongming Eco-Island in Shanghai, there still remain some changes to identify and assess what an eco- planning should represent. We refer back to last year's visit of governmental authorities to the site and putting on halt any of the high-rise development projects.

In most of our studied cases at the meso and macro scales in China, we identify a high potential—and still not fully exploited—for integrated design and planning. This would then allow for more opportunities of eco-innovation, not only to be recognised as development of final products but also processes that support the ideals of sustainable development. Some common examples can be related to environmental improvements and technological advancements, as mentioned in our cases of potential future directions. One method towards achieving advanced thinking would then propose for the integration of multiple dimensions, treating the built environment not necessarily as an information-based model, but rather as a platform for comprehensive assessment and ultimately towards holistic implementation. The linkage between innovation and sustainability is emerged as the concept of 'eco-innovation', which was first discussed by Claude Fussler and Peter James (1996). Since then, the term is recognised as a breakthrough discipline for both sustainability and innovation. The later definitions of the concept focused on key factors of sustainability, such as reduction in environmental impact (James 1997) and tools of change towards sustainable development (Rennings 2000). In theory, it can easily be translated as part of eco-development thinking, as it clearly defines the goals of environmental sustainability, technological advancements, eco-planning, eco-design, eco-efficiency, environmental technology, environmental design and sustainable innovation. Nevertheless, the only way forward is to consider how this can be integrated at multiple scales and with the consideration of multiple dimensions. In this case, we suggested for a new integrated model of '*Eco-Fusion*' as an advanced urban sustainability method towards eco development.

Coined by Cheshmehzangi (2016), the term 'eco-fusion' is a representation of integrating eco-development strategies across three spatial levels of the built environment. This approach is a comprehensive tool to achieving eco-development in multiple levels. A tool that offers a blend of eco-development strategies and possibilities that may eventually enhance the current practice of eco-design and planning, comprising three levels of eco-design at three urban spatial levels of eco-city (city/district level), eco-community (neighbourhood level) and eco-building (site/building level). This enables us to consider eco-development not only through its horizontal indicators and dimensions, but also through its verticality of scale, implementation, and governance. In other words, the eco-strategies can vertically run through our spatial levels, creating a more comprehensive process of city management and integration in planning and design.

The case of Sino-Singaporean Tianjin Eco-City partially achieves this by linking the ideals of eco-planning into eco-community design and green-certified buildings. Furthermore, this enables to create a more transparent and effective framework for stakeholder constellations of different levels, e.g., from city/municipal level management level, to district level, and to communities and site management. In this process, we anticipate to see a better information sharing for project data, a more comprehensive management network, and a more enhanced dialogue between the engaged stakeholders. The concept of eco-fusion is purely an approach to consolidate eco-strategies in both horizontal and vertical directions, to enable a better management for governance and adaptability for innovation.

Further to our discussion, we urge to avoid mere 'branding' of the eco-development ideals (Cheshmehzangi 2017) and instead focusing on the application of eco-thinking in practice. This enables us to focus on not fulfilling the eco-branding but rather the methods in which eco- can help us to achieve sustainable environments. To elaborate, we can propose for future hybrids, such as eco-low-carbon city, smart-eco-city, and smart-resilient city (ibid.), not because it seems necessary to create a new branding model, but instead because it is essential to think holistically and considering any possible tool or combination of a few to achieve sustainable development.

REFERENCES

Agyeman, J., McLaren, D., Schaefer-Borrego, A. (2013) Sharing cities. Friends of the Earth.

Ahmad, S., de Oliveira, J.A.P. (2016) Determinants of urban mobility in India: Lessons for promoting sustainable and inclusive urban transportation in developing countries. Transport Policy 50, 106–114.

Bansal, P., and McKnight, B. (2009) Looking forward, pushing back and peering sideways: Analyzing the sustainability of industrial symbiosis, *Journal of Supply Chain Management*, 45(4), pp. 26–37.

Belk, R. (2014) You are what you can access: Sharing and collaborative consumption online. Journal of Business Research 67, 1595–1600.

Bin, S., and Nadel, S. (2012) *How Does China Achieve a 95% Compliance Rate for Building Energy Codes? ACEEE Summer Study on Energy Efficiency in Buildings 2012*, available at: http://aceee.org/files/proceedings/2012/data/papers/0193-000261.pdf.

Cao, D., Li, H., Wang, G., Huang, T. (2017) Identifying and contextualising the motivations for BIM implementation in construction projects: An empirical study in China, *International Journal of Project Management*, Volume 35, Issue 4, pp. 658–669.

Chen, B., (2014) Green building concepts in Chinese vernacular buildings (浅谈古建筑中的绿色建筑理念), Doors and Windows (门窗), (6): 202–202.

Chen, J., and Wang, J. (2012) How traditional Chinese architecture culture affects green building design (中国传统建筑文化对绿色建筑的影响探讨), Green Technologies (绿色科技), (8): 272–274.

Cheshmehzangi, A. (2014) The Urban and Urbanism: China's New Urbanization and Approaches towards Comprehensive Development, *The International Journal of Interdisciplinary Environmental Studies*, Vol. 8, Issue 3–4, pp. 1–12.

Cheshmehzangi, A. (2016) *Eco Fusion: A New Approach to Multi-Spatial Eco Development*, Technical Report, Published in Ningbo, China.

Cheshmehzangi, A. (2017) '*Sustainable Development Directions in China: Smart, Eco, or Both?*' in the 2017 International Symposium on Sustainable Smart Eco-city Planning and Development, 11–13 Oct 2017 Cardiff, UK.

Cheshmehzangi, A., and Butters, C. (eds.) (2017) *Designing Cooler Cities – Energy, Cooling and Urban Form: The Asian Perspective*, London: Palgrave Macmillan.

Cheshmehzangi, A., Deng, W., Zhang, Y. and Xie, L. (2017a) *Green Development Status in Zhejiang Province and the City of Ningbo, China: Examination of Policies, Strategies and Incentives at Multiple Levels*, IOP Conference Series Earth and Environmental Science, Vol. 63(1), pp. 1–7.

Cheshmehzangi, A., Xie, L. and Tan-Mullins, M. (2018) The Role of International Actors in Low-Carbon Transitions of Shenzhen's International Low Carbon City in China, *Journal of Cities*, Published November 2017.

Cheshmehzangi, A., Zhu, Y., and Li, B. (2017b) Application of environmental performance analysis for urban design with Computational Fluid Dynamics (CFD) and EcoTect tools: The case of Cao Fei Dian eco-city, China, *International Journal of Sustainable Built Environment*.

China Society of Urban Studies (2016) *China Green Building Annual Report*. Beijing: China Building Industry Press.

Cohen-Shacham, E., Walters, G., Janzen, C., and Maginnis, S. (eds) (2016) *Nature-based solutions to address global societal challenges*. Gland, Switzerland: IUCN. Xiii + 97 pp. Available at: https://portals.iucn.org/library/node/46191.

Cole R.J. (2010) *Environmental assessment: shifting scales*, In Edward Ng (eds) Designing high-density cities for social and environmental sustainability, pp. 273–282, Earthscan London.

Coombes, E., Jones, A.P. and Hillsdon, M (2010) The relationship of physical activity and overweight to objectively measured green space accessibility and use. *Social science & medicine* 70, 816–822.

Coutts A., and Harris R. (2013) *A multi-scale assessment of urban heating in Melbourne during an extreme heat event: policy approaches for adaptation*; Available at: http://www.vcccar.org.au/sites/default/files/publications/Multiscale%20assessment%20urban%20heating%20Technical%20Report.pdf.

Cunningham, R., Christopher C., Thomas M., Brent J., Anne-Maree, D. and Ben Harman (2016) Engaging communities in climate adaptation: the potential of social networks. *Journal of Climate Policy*, Volume 16, Issue 7.

Deng, W., Cheshmehzangi A., and Yang T. (2014) Environmental implications of privatised public space in gated residential neighbourhood, *International Journal of Social Science and Humanity*, (IJSSH, ISSN: 2010-3646).

Deng, W; Blair, J; Yenneti, K (2017) *Contemporary urbanization: challenges, future trends and measuring progress*, in Encyclopedia of Sustainable Technologies, edited by Abraham M. et al. Elsevier Oxford UK.

Desouza, K.C. and Flanery, T. H (2013) Designing, planning, and managing resilient cities: A conceptual framework. *Cities*, 35, pp. 89–99.

EURObiz (2013) *Sustainable urbanisation drives innovation, Published by Journal of European Union Chamber of Commerce in China*. Available at: http://www.eurobiz.com.cn/sustainable-development-urbanisation-drives-innovation/.

Faivre, N., Fritz, M., Freitas, T., de Boissezon, B., and Vandewoestijne, S. (2017) Nature-Based Solutions in the EU: Innovating with nature to address social, economic and environmental challenges, *Environmental Research*, Vol. 159, pp. 509–518.

Fan, P., Ouyang, Z., Basnou, C., Pino, J., Park, H., Chen, J. (2017) Nature-based solutions for urban landscapes under post-industrialization and globalization: Barcelona versus Shanghai, *Environmental Research*, Volume 156, pp. 272–283.

Feng, K., Fridley, K., and Huang, Z. (2016) *Impact Analysis of Developing Net Zero Energy Buildings in China. ACEEE Summer Study on Energy Efficiency in Buildings*. Available at: http://aceee.org/files/proceedings/2016/data/papers/10_1135.pdf.

Fussler, C. and James, P. (1996) *Driving Eco-Innovation: A Breakthrough Discipline for Innovation and Sustainability*, Pitman Publishing: London.

Gonela, V., Zhang, J. and Osmani, A. (2015) Stochastic optimization of sustainable industrial symbiosis based hybrid generation bioethanol supply chains, *Computers & Industrial Engineering*, vol. 87, pp. 40–65.

Habibi, S. (2017) The promise of BIM for improving building performance, *Energy and Buildings*, Volume 153, pp. 525–548.

He B.J., Yang L., and Ye M. (2014) Building energy efficiency in China rural areas: Situation, drawbacks, challenges, corresponding measures and policies. *Sustainable Cities and Society* 11:7–15.

Hodson, M. and Marvin, S. (2010) Urbanism in the Anthropocene: Ecological urbanism or premium ecological enclaves? *City*, Vol. 14, Issue, 3, pp. 298–313.

James, P. (1997) The Sustainability Circle: a new tool for product development and design, *Journal of Sustainable Product Design*, Vol. 2, pp. 52–57.

Joss, S. (2011) Eco-Cities: The Mainstreaming of Urban Sustainability; Key Characteristics and Driving Factors, *International Journal of Sustainable Development and Planning*, 6:3, pp. 268–285.

Keesstra, S., Nunes, J., Novara, A., Finger, D., Avela, D., Kalantari, Z., and Cerdà, A. (2018) The superior effect of nature based solutions in land management for enhancing ecosystem services, *Science of The Total Environment*, Volumes 610-611, pp. 997–1009.

Kelly, C., Essex, S., and Glegg, G. (2012) Reflective practice for marine planning: A case study of marine nature-based tourism partnerships, *Marine Policy*, Volume 36, Issue 3, pp. 769–781.

Li, J. (2011) Decoupling urban transport from GHG emissions in Indian cities—A critical review and perspectives. Energy policy 39, 3503–3514.

Lovell, H. (2012) *Eco-Communities*, International Encyclopedia of Housing and Home. Elsevier Publication, pp. 1–5.

Mytton, O.T., Townsend, N., Rutter, H. and Foster, C (2012) Green space and physical activity: an observational study using health survey for England data. Health & place 18, 1034–1041.

Nikolaidis, N. P., Kolokotsa, D., and Banwart, S. A. (2017) Nature-based solutions: business, *Nature*, 543 (7645), pp. 315–315.

Norman, B. (2016) Climate Ready Cities. Policy Information Brief 2, National Climate Change Adaptation Research Facility, Gold Coast.

Panno, A., Carrus, G. Lafortezza, R., Mariani, L., and Sanesi, G. (2017) Nature-based solutions to promote human resilience and wellbeing in cities during increasingly hot summers, *Environmental Research*, Volume 159, pp. 249–256.

Qiu, X., and Huang, G. Q. (2013) Supply Hub in Industrial Park (SHIP): The value of freight consolidation, *Computers & Industrial Engineering*, Vol. 65, pp. 16–27.

Rennings, K. (2000) Redefining innovation – eco-innovation research and the contribution from ecological economics, *Ecological Economics*, Vol. 32 (2), pp. 319–332.

Rifkin, J. (2014) *The Zero Marginal Cost Society: The Internet of Things, The Collaborative Commons, And The Eclipse of Capitalism*, New York: Palgrave Macmillan.

Roseland, M. (1997) Dimensions of the eco-city. *Cities*, 14, 197–202.

Saleh, F., and Weinstein, M. P. (2016) The role of Nature-based Infrastructure (NBI) in Coastal Resiliency Planning: A Literature Review, Journal of Environmental Management, Volume 183, Part 3, pp. 1088–1098.

Santamouris, M. (2015) Regulating the damaged thermostat of the cities – status, impacts and mitigation challenges. Environmental building.

Shahrokni, H., Levihn, F., and Brandt, N. (2014) Big meter data analysis of the energy efficiency potential in Stockholm's building stock, *Energy and Buildings*, vol. 78, pp. 153–164.

Shi, Q., Yu, T., Zuo, J., and Lai, X. (2016) Challenges of developing sustainable neighbourhoods in China, *Journal of Cleaner Production*, Vol. 135, pp. 972–983.

SOHO BIM project report (2013) *internal project report on SOHO Building*, Shanghai, unpublished document.

Theeuwes, N.E., Steeneveld, G.J., Ronda, R.J., Heusinkveld, B.G., van Hove, L.W.A. and Holtslag, A.A.M. (2014) Seasonal Dependence of the Urban Heat Island on the Street Canyon Aspect Ratio, Quarterly Journal of the Royal Meteorological Society, Volume 140, Issue 684 October 2014 Part A Pages 2197–2210.

van den Bosch, M. and Ode Sang, A. (2017) Urban natural environments as nature-based solutions for improved public health – A systematic review of reviews, *Environmental Research*, Volume 158, pp. 373–384.

van der Nat, A., Vellinga, P., Leemans, R., and van Slobbe, E. (2016) Ranking coastal flood protection design from engineered to nature-based, *Ecological Engineering*, Volume 87, pp. 80–90.

Vance, C., Hedel, R. (2007) The impact of urban form on automobile travel: disentangling causation from correlation. Transportation 34, 575–588.

Vujcic, M., Tomicevic-Dubljevic, J., Grbic, M., DusicaLecic-Tosevski, D., Vukovic, O. and Toskovic, O. (2017) Nature based solution for improving mental health and well-being in urban areas, Environmental Research, Volume 158, pp. 385–392.

Weber, R. and Reardon, M. (2015) Do eco-districts support the regional growth of cleantech firms? Notes from Stockholm, *Cities*, Vol. 49, pp. 113–120.

Wild, T. C., Hennberry, J., and Gill, L. (2017) Comprehending the multiple 'values' of green infrastructure – Valuing *nature-based solutions* for urban water management from multiple perspectives, *Environmental Research*, Volume 158, pp. 179–187.

World Economic Forum (2016) Shaping the future of construction – a breakthrough in mind-set and technology, prepared in collaboration with the Boston Consulting Global, May 2016.

Wu, Y., J. Hou, N. Zhou and W. Feng. (2015) *Research on an Energy-Efficiency Improvement Roadmap for Commercial Buildings in China*. Proceedings of the European Council for An Energy- Efficient Economy 2015 Summer Study on Energy Efficiency. Club Belambra Les Criques, Presqu'île de Giens Toulon/Hyères, France: ECEEE.

Xiao, Z., Cao, B., Zhou, G. and Sun, J. (2017) The monitoring and research of unstable locations in eco-industrial networks, *Computers & Industrial Engineering*, Vol. 105, pp. 234–246.

Yigitcanlar, T. and Lee, S. H. (2014) Korean ubiquitous-eco-city: A smart-sustainable urban form or a branding hoax? *Technological Forecasting and Social Change*, 89, pp. 100–114.

Xu, W. (2017) Nearly zero energy building research and development in China (中国近零能耗建筑研究和实践). Science & Technology Review (科技导报), 2017, 35(10): 38–43.

Yang, Z., Fan, Y., and Zheng, S. (2016) Determinants of household carbon emissions: Pathway toward eco-community in Beijing, *Habitat International*, Volume 57, pp. 175–186.

Zhang B., Min S.G., Yang T., and Yang B. F. (2014) Li-fang system, urban neighbourhood and green city 里坊城市●街坊城市●绿色城市, China Architecture & Building Press (in Chinese).

Zhang, D. (2011) *Study on Promoting Government Mechanism of Public Participation In Low-carbon Community Construction*. Unpublished Doctoral Dissertation. Wuhan University of Science and Technology, Wuhan.

8.4.6 Websites

3ders.org website (2015) http://www.3ders.org/articles/20150118-winsun-builds-world-first-3d-printed-villa-and-tallest-3d-printed-building-in-china.html.

China Daily (2015) 3D printing changes house building; Available at: http://www.chinadaily.com.cn/china/2015-01/20/content_19353033.htm.

China Daily (2016) Gated community will open gradually. Available at: http://www.chinadaily.com.cn/china/2016-02/24/content_23624204.htm.

New Atlas (2014) Chinese company built 10 houses in a day. Available at: https://newatlas.com/china-winsun-3d-printed-house/31757/.

China Property (2017) Low Compliance Rate in Building Energy efficiency, newspaper report at 2014.04.01, http://house.ifeng.com/detail/2014_04_01/45359169_0.shtml 中国房地产报, 建筑三步节能达标率低, 亟待推行节能测试.

Khalamayzer, A. (2016) Sustainability reporting comes of age. Available at https://www.greenbiz.com/article/sustainability-reporting-stock-exchanges-comes-age?utm_source=newsletter&utm_medium=email&utm_term=newsletter-type-greenbuzz-daily&utm_content=2016-12-11&utm_campaign=newsletter-type-greenbuzz-daily-106566&mkt_tok=eyJpIjoi-WVdNelkyWmhOelJtTURJdyIsInQiOiI0VHVuMUlFb3YzSDdvXC9n-b3VjUTMzVG1HMWczTDgzbDBsOUhaaTNEeGo4bXF1eHNuQ0IzcTc-wXC9lNmtHbk14M1o5RUh4dlwvc1YwN2R5TGY0K3hrQ2FhVVh3T-GRcL1lDMEdlNVZZYWlrXC9qbFo3cm9ZdTBuWThIQWIyVkJsMnVWc3BNIn0%3D.

National Geography article (2017) Available at: https://news.nationalgeographic.com/2017/03/china-takes-leadership-climate-change-trump-clean-power-plan-paris-agreement/.

Winsun website. Available at https://futureofconstruction.org/case/winsun/.

CHAPTER 9

Concluding Remarks

9.1 Recapitulation

There is a debate that the recent nationwide environmental campaign, which began in early 2017, may have a negative impact on economic growth and employment. This campaign has led to closure of many factories that violated environmental protection laws and regulations. Li Ganjie, China's minister of Environmental Protection, refuted this claim at a press conference of the 19th National Congress of the party held between 18 and 24 October 2017. Instead of reducing economic development, environmental protection and green development, as the minister emphasised, will further consolidate the economic growth foundation and achieve a much more sustainable development (China Daily 1, October 2017). Addressed by President Xi in the report to the Party Congress, increasing harmony between humans and nature is one of the 14 basic policies underpinning endeavours to develop socialism with Chinese characteristics in the new era (China Daily 2, October 2017). It is expected that current policies, strategies and pilot projects, at both national and local levels, will further address urban environmental constraints. They are expected to be further enhanced in the forthcoming years. One important purpose is to identify an urban development model, which can be replicable in a wide range of contexts in the country. More importantly, however, such a development model can refine a balance between economic development and environmental protection; in other words, the 'harmonious development' that is anticipated.

© The Author(s) 2018
W. Deng, A. Cheshmehzangi, *Eco-development in China*,
Palgrave Series in Asia and Pacific Studies,
https://doi.org/10.1007/978-981-10-8345-7_9

This book explores contemporary eco-development projects in the context of China. These project cases were categorised and assessed at three interwoven spatial levels: city; community/neighbourhood; and buildings. A small selection of 21 successful projects at different spatial scales were examined in the earlier chapters (see Chaps. 5, 6, and 7). These case studies provide international readers with a comprehensive and timely picture of current eco-development practices in China. And many of these cases are unfamiliar to readers outside China. These cases also offer a variety of eco-development approaches and strategies that have shaped the first phase of eco- and green projects in China. They are developed in a relatively short timeframe, within which China has taken many initiatives to learn from international examples, adapt and develop new development strategies, and took steps to policy reforms.

Following an evaluation of these case study projects, we then looked into the past and current trends of eco-developments in China. We further argued about the importance of integration in eco-development strategies and planning. This can only occur by consideration of the multiscalar and multidimensional nature of urban sustainability which provides us with models of best practice. We refer to this integration as the interplay between spatial levels of the built environment. It is, therefore, highly important to highlight both the existing gaps and the potential relationships of spatial levels. Moreover, the current construction industry is also being shaped by emerging new technologies, which are able to increase urban efficiency and performance. However, the adoption of new technologies in the construction sector is still relatively slow globally, as suggested by a recent report from the World Economic Forum (2016). In contrast, China has been the largest construction market over the past three decades, which provides an 'experimental platform' for various new urban and building approaches. This experimental phase may have not been very innovative at first, but it has certainly matured in recent times, and particularly in the last two to three years. Furthermore, it is already evident that China has become a global pioneer in eco-city endeavour and that many new technologies are being used experimentally to increase urban and building performance; such as urban-sharing services and 3D printing technology for construction. More emerging technologies are tested and implemented in small-scale projects, providing the opportunity for scaling up cases and success stories that may become widespread. Some of these projects start at a very small scale of micro or meso scale, and are now visible at macro and event larger scales (e.g., the example of the bicycle-sharing system that has grown from a university campus idea to an international market). In the concluding chapter, we further explore the concept of eco-development in China with a comprehensive strengths, weaknesses, opportunities and threats (SWOT) analysis that highlights what China faces in the current stage of development. These will refer back to some of the ideals of the future that we highlighted in the discussion chapter (see Chap. 8). In this section, we also raise some questions of interest for future research on the matters of sustainable city, sustainable community and green building development in China.

9.2 SWOT ANALYSIS

9.2.1 Strengths

By promoting eco-city projects in recent years, China appears in the front line of reshaping and redeveloping the urban environments. China has initiated new policies, strategies and pilot projects at both national and local levels to address resource and environmental constraints in Chinese cities. These criteria are continuously highlighted in China's five-year plans (i.e., particularly in the 11th, 12th, and 13th FYPs, so far). Conventionally, sustainable city efforts have been focused on individual issues such as urban energy, urban transport, land use, waste, water and urban health. In recent years, there is a growing interest in China to find a new integrated model for urban development as a whole; a method that we define as a comprehensive urban development model. Efforts to create such a new urban development model is manifested in the form of developing eco-cities. In this approach, the emphasis remains on the role of interplay between spatial levels.

Although it is still too early to judge the success of the eco-city concept in new urban development areas in China (i.e., as many projects are still under construction), it is very likely to see significant influence on the planning, design and operation of cities in the future. For example, in the Sino-Singaporean Tianjin Eco-City (SSTEC), the implementation of the eco-city concept has shaped the local construction industry and green building development, from the perspectives of design, material manufacturing and construction through to operational management. Eco-development has also become a valuable contributor to the local economy. Some cases have already offered experimental planning scenarios, some of which that can simply be adapted in any future projects (e.g., the concept of eco-cell in the SSTEC project, see Sect. 5.6).

One significant impact is on the local government to improve urban performance through various methods of integrated planning, design, and management approaches. This is particularly important in the coastal cities in the eastern and southern parts of the country, where the majority of large-scale cities are located. In many cases, the local government has started to look into new ways of urban governance to achieve eco-development (also identified as green development). One example is the New Eastern City in Ningbo Zhejiang Province, which is a new expanded area from the existing city centre with 16 km^2 and is planned to accommodate a population of 170,000 people, including both the existing population and new residents. In the original development plan, the new city was developed with two phases—completing the western part of the city between 2006 and 2014, and the eastern part between 2015 and 2020. The western part was completed as planned in 2014, which was developed with conventional planning and design control criteria, such as building heights, landscape features and infrastructures. Very little considerations were related to eco-development at that time. As eco-development has escalated in recent years and exemplary projects have been widely known, the local government approached the University of Nottingham Ningbo China (UNNC) in 2015 and funded a research project, led by the two authors of this book. The project was initiated with the sole intention of improving ecological performance of the eastern part of the city through the adoption of a new system for urban governance. This also highlights the role of interplay between two of the spatial levels, the meso and micro levels, within which new eco-strategies are proposed and planned for further implementation. The key requirements of this project are:

- Develop a sustainability indicator system for the city as a whole;
- Break down these overall sustainability indicators to all plots of land that will be sold to individual developers through an open bidding process. These plot-level requirements will be incorporated into the tendering document as new planning and design constraints, along with those conventional criteria;
- Develop a technical guidance for various building typologies (e.g., schools, hospitals, offices and residential houses) to achieve green building ratings.

This is the first time a comprehensive system is developed to promote eco-development in the city of Ningbo as part of its urban governance. Based on the current status and importance of the project, it will have a profound impact on the city's urban development in future. Therefore, we can also potentially see an interplay between the meso and macro levels. A large group of stakeholders from the government, developers and practitioners are incentivised to change their 'business-as-usual' mode of dealing with urban development by considering these new requirements. This already is a step forward, as it enables a variety of stakeholder groups to at least think in a different direction to the current 'business-as-usual' scenarios. This research also provides a new methodology to break down the city-level overall eco-targets to neighbourhood-level (land plot) indicators, and then building-level (building site plot) sub-indicators. Once again, this highlights the importance of interplay between the spatial levels, where we can create potential collaborations between various stakeholder groups of the project. In this project, the key issues considered comprise green building ratings, energy consumption, renewable energies, pre-fabricated building elements and water recycling. All of these issues are penetrated through city, neighbourhood and building levels, and correspond at different spatial levels. By doing this, the three interrelated spatial levels are able to be examined in the same dimension and form a collective effort towards a comprehensive eco-development model.

9.2.2 Weaknesses

China has advanced to a higher level of eco-development through the adoption of new emerging technologies. These new technologies, such as the Internet of Things (IoT), smart cities and 'Big Data', Artificial Intelligence (AI) and sharing services, are able to increase urban efficiency in a variety of ways. However, they also bring new challenges to urban governance in China, which has lagged behind in terms of accommodating these 'changes' or 'transitions' brought about by new technologies. The combined method of *eco- strategies* and smart initiatives is growing rapidly and is already becoming a method of promoting urban innovation in many Chinese cities.

One example of sharing services is the growing dockless shared bike scheme that has flooded the streets of many Chinese cities in the last few years. This initiative is already expanding to other countries, such as Australia and the UK. For instance, since April 2016, just one of the shared bike operators, MoBike, has placed more than 100,000 bikes in Shanghai, Beijing, Shenzhen and Guangzhou (The Guardian 2017). The way the schemes work is simple—users download an app that informs them about the location of bikes, which they can unlock by scanning a QR code by their phones or using a combination code they have received. Unlike traditional rental services, which require bikes to be returned to a fixed docking station, riders are free to leave the bikes wherever their journey ends. Users pay for using the bikes through online banking systems (ibid.).

The bike-sharing programmes have enabled bikes to return to the streets of China, especially in big cities where private vehicles are the primary means of transportation. However, the rapid expansion in the use of dockless bikes has brought new challenges in terms of urban governance. Users park the bikes wherever they want, causing congestion in pedestrian walking paths and other public spaces, which is considered detrimental to the image of the city. Consequently, some city governments have suspended the growth of shared bikes and many neighbourhoods and public spaces have banned shared bikes from entering (shown in Fig. 9.1). This impairs the use of shared bikes as an efficient means of transport over short distance in Chinese cities. To accommodate these challenges, urban governance in areas such as spatial planning and parking management should be upgraded accordingly (Fig. 9.1).

Fig. 9.1 The view of several shared bikes from various companies left outside the entrance area of the University of Nottingham Ningbo China. These bikes have now been banned from the university campus area (photo taken by the authors)

Another challenge lies in the ways of providing indoor environmental quality (IEQ) within buildings, including thermal, acoustic and lighting comforts, along with indoor air quality. It is recognised that the two main disciplinary perspectives in the IEQ domain are passive approaches and active approaches. To date, building developers and designers in China have opted to adopt more such 'active' measures to achieve higher performance. Equally important, however, is the passive approach to improve indoor environmental quality, with its focus on low-energy and low-cost design strategies. The two approaches can be applied singularly or in carefully-balanced combinations that are appropriate to the external environmental and climatic contexts. However, the passive approach should be maximised before the active approach is utilised.

9.2.3 Opportunities

As discussed earlier, the sustainable built environment is comprised of multiple dimensions (environment, society, economy, and governance). It is also viewed across the lifecycle, and across dozens of subject areas from material manufacturing to building design and engineering, and to indoor environmental quality, community cohesion and urban planning. The foregoing discussions explore the contemporary sustainable built environment research and practice at the three interwoven themes of city, community and individual building—each of them have their specific problems to deal with.

Sustainability efforts will be less effective at the other spatial levels if a spatial level is not properly addressed. For example, the spatial arrangements of neighbourhoods have significant impacts on the environmental performance of buildings and impose direct, as well as indirect, costs on households. Another example is the formation of the urban heat island effect, which is largely affected by the design choices of individual buildings. A summary of pathways to sustainable built environment at the three spatial levels is listed in Table 9.1.

While each of the spatial levels is coupled intrinsically with specific sustainability requirements and solutions, a potentially greater set of performance gains lie in synergies between the interplay of the three spatial levels. In this book, we argue in favour of an effective urban governance system that should be established to integrate the efforts at the three spatial levels of the built environment. The key is to develop a set of urban key performance targets and break them down to different spatial levels

Table 9.1 Summary of pathways to sustainable built environment from the consideration of all three spatial levels in the built environment

Knowledge areas	Pathways to sustainable built environment
City level	Policies, master planning, ecology, transportation, wastes, air pollution, and etc.
Neighbourhood/community level	Education, behavioural patterns, open space, health, neighbourhood pattern, walkability, and etc.
Building level	Performance, rating tools, architectural design, indoor comfort, IT technologies, renewable energies, and etc.

(Fig. 9.2). This can provide a common ground to monitor the progress at different levels and form a collective effort to move a city towards sustainability and sustainable development.

9.2.4 Threats

The current eco-city venture have not led to a definitive definition of eco-city as no agreeable definition has emerged to date (de Jong 2015). This is particularly complicated in the context of China, where eco-city projects vary significantly in terms of their size, purpose, development modes and growth. Joss (2015) addresses the plausible reasons for this situation. Firstly, it is suggested that diversity and ideas set in specific locales and are governed by traditions. In addition to this, the competition between various actors has led to a lack of cohesion regarding the vision and roadmap to eco-city development. Secondly, this lack of coherence is simply an inevitable process towards consensus, as the current phase could be seen as the 'experimental phase'. Under such a reading it is anticipated that this phase is then followed by a phase of consolidation, leading to more agreeable international norms. Nonetheless, de Jong (2015), who also identifies the terminological fuzziness and confusion, stresses the need for clarification into distinct categories in relation to specific applications. This argument invariably endorses the context dependency of eco-city and its implementation of its indicators.

Our earlier examination of the key performance indicators (KPI) systems of eight eco-cities in China indicated what has been considered in the planning system as a potential mean to achieve sustainable eco-development (Deng et al. 2017). However, these KPI systems vary greatly in assessment focuses and priorities. This further validates the lack of eco-city definition

Fig. 9.2 Breakdown of urban key performance targets from urban level to community and building levels (image drawn by the authors)

across Chinese projects. Such confusion may lead to a conceptually vacuous principle, i.e., lacking content with a threat of its application being applied for promotional and market-fashionable services. On the other hand, there are still some convergences from the issues commonly addressed by the reviewed KPI systems (ibid.). These issues are mainly environment-related, concerning matters such as energy, water and waste management. We argue that an agreement on eco-city development targets should be established, although the approaches to achieve the targets may vary depending on specific context and its attributes for new development, future growth and expansion. Therefore, future eco-city practices need to address these common targets in the local contexts.

9.3 Final Remarks

This book is the first attempt to highlight and assess a variety of eco-development projects at multiple spatial levels of the built environment in China. It also stresses the importance of interplay between these spatial levels, by addressing the gaps, issues and challenges at each level, and also between levels. In this book, we have argued that strategies and solutions for eco-development must consider the interplay between various spatial levels, minimise policy conflicts, and support innovative directions of holistic planning and design. Eco-development is highly influenced by policy development, and we urge the need to accelerate policy reforms and transitional cases, which enable some of the existing cities to consider eco- as their main agenda of future development, enhancement and growth. Future eco-development cases may require the inclusion of more comprehensive system thinking approaches, and may also highlight the importance of ecology and nature in the built environment. The future projects should also look into key factors of transitions, such as energy transitions and low-carbon transitions, and find methods of achieving the green economy. Through this thinking, we require to adapt new strategies that enable us to broaden our profitability across various dimensions of sustainability and for various stakeholder constellations. By doing so, we will be in the right direction to change the current behaviours, mindsets and preferences of not only the users, but those of decision makers, governmental authorities, policy makers, developers and the practitioners.

Although we focused on cases, strategies and solutions in China, the ideas that have emerged from this book should be applicable to any other country with a strong green or eco-development agenda. Also in the contexts were institutions for a greener development are weak, or directions are not necessarily eco-friendly, we hope to see immediate changes and new directions other than 'business-as-usual' practices. As it stands, China will set new practices, some of which may become 'best practices', and some of which may become 'new development modes' that may be applied in other regions. In this process, what will remain visible is the growing role of China in this direction of eco-development. In this respect, we anticipate seeing new trends of development, new planning models and new strategies towards the future of eco-development.

References

China Daily 1 (Oct 2017) Environmental Protection Won't Slow Down Economic Growth. Available at: http://www.chinadaily.com.cn/china/19thcpcnationalcongress/2017-10/23/content_33613817.htm.
China Daily 2 (Oct 2017) Highlights of Xi's Report to the Congress. Available at: http://www.chinadaily.com.cn/china/2017-10/19/content_33436874.htm.

de Jong, M., Joss, S., Schraven, D., Zhan C., and Weijnen, M. (2015) Sustainable–smart–resilient–low carbon–eco–knowledge cities; making sense of a multitude of concepts promoting sustainable urbanization, *Journal of Cleaner Production*, 109, pp. 25–38.

Deng, W., Cheshmehzangi, A., Dawodu, A., and Wang, B.Y. (2017) *Comparison study of eco-city key performance indicator systems in China*, Proceedings of World Conference for Sustainable Built Environment 2017, pp. 791–802, 5–7 June 2017, Hong Kong.

Joss, S. (2015) *Sustainable Cities: Governing for Urban Innovation*, London: Palgrave Macmillan.

The Guardian (2017) Uber for bikes: how 'dockless' cycles flooded China – and are heading overseas; Available at: https://www.theguardian.com/cities/2017/mar/22/bike-wars-dockless-china-millions-bicycles-hangzhou.

World Economic Forum (2016) Shaping the future of construction – a breakthrough in mind-set and technology, prepared in collaboration with the Boston Consulting Global, May 2016.

Index

A
Aesthetics, 33, 180, 191, 198–200
Agenda 21, 17, 23, 73
Air conditioning, 61, 124, 192, 194, 206, 207, 209
Air pollution, 20, 92, 97, 119
Ancient China, 87, 232–234
Ancient Chinese architecture, 233
Anyang, 2–4, 19
Artificial Intelligence (AI), 239, 259
Ascendas OneHub Business Park, 171–173
Asia, 1, 23, 24, 83

B
Behavioural change, 220, 222
Beijing, 96, 98, 120, 131, 133, 167–170, 176–180, 188, 189, 198–200, 234, 259
Beijing Grand MOMA, 178–180
Beijing Olympic Village (BOV), 176–178
Benchmark, 58, 60, 87, 89, 91, 96, 122, 222
Best practice, 18, 36, 46, 51, 54, 113, 136, 165, 171, 216, 230, 247, 256, 264
Big Data, 45, 119, 120, 239, 259
Binghai Xiaowai Primary School, 208–209
Biodiversity, 4, 17, 18, 20, 23, 52, 55, 113, 245
Black water treatment, 61, 122
Bruck residence, 195–198
Brundtland Report, 14, 15, 17
Building code, 223, 225
Building efficiency, 60
Building energy consumption, 54, 87, 90, 187, 195, 209, 210, 223, 225, 233
Building envelope, 29, 197, 199, 202
Building form, 192
Building information modelling (BIM), 101, 240–242
Building orientation, 199, 243

© The Author(s) 2018
W. Deng, A. Cheshmehzangi, *Eco-development in China*,
Palgrave Series in Asia and Pacific Studies,
https://doi.org/10.1007/978-981-10-8345-7

Building Research Establishment
    Environmental Assessment
    Method (BREEAM), 31, 51, 60,
    67, 95, 191, 216
  Communities, 68–70, 158, 159

C
Carbon footprint, 172, 174, 178, 180,
    192, 195, 200
Cave dwelling, 234–236
Center for Sustainable Energy
    Technologies (CSET), 190–195
Central Business District (CBD),
    122, 144, 160, 173–175, 226
Changsha, 95, 105, 120
Chenjia Town, 108, 111, 112
China's urbanisation, 8–10, 17,
    81–86, 133, 215, 227
Chinese patio house, 234–235
Chongming Eco-Island, 101,
    106–115, 151, 247
Climate adaptation, 246
Climate change, 3–5, 18, 20,
    21, 23, 27, 35, 71, 73,
    74, 89, 217, 226–227,
    238, 239, 247
Cloud computing, 119, 120, 160
Community cohesion, 31, 33,
    39, 261
Compact city, 34, 52, 70, 92, 233
Comprehensive Assessment System for
    Built Environment Efficiency
    (CASBEE), 65, 67
Construction management, 70
Cooling effect, 124
Cooling energy, 61
Courtyard, 59, 199, 233, 234

D
District energy system, 221

E
Earth system, 5
Eco-business park, 158, 171, 172,
    173, 180
Eco-cell, 134, 141, 222, 257
Eco-footprint (EF), 4, 18, 19, 58,
    228, 229
e-Footprint, 245
Eco-fusion, 247–249
Eco-industrial park, 158, 159–160,
    180, 221, 221, 230
Eco initiatives, 160–161, 163–172,
    174, 176–177, 179, 192–209
Eco-innovation, 221, 222, 247–249
Ecological civilisation, 115–118, 132,
    162, 217
Ecological impact, 228
Eco-residential compound, 158, 159,
    176, 178–179, 180
Eco-village, 23, 57, 158, 164–165,
    168, 180, 230
Electricity, 26, 33, 58, 59, 97, 113,
    139, 166, 177, 204
Electricity consumption, 58
Embodied energy, 55, 190, 244
Energy consumption, 18, 20, 29, 54,
    87, 89–91, 93, 97, 122, 124, 161,
    179, 187, 190, 195, 203, 209,
    210, 222–225, 233, 243, 259
Energy efficiency, 31, 55, 57, 87, 89,
    91, 97, 100, 113, 135, 140, 163,
    169, 177, 199–201, 222–225,
    228, 238, 244
Energy efficient buildings, 54, 55,
    148, 168, 195, 196, 207
Energy management, 121, 140, 163,
    179, 194
Energy Performance of Buildings
    Directive (EPBD), 243
Energy production, 61, 148
Energy source
    passive, 179, 194

renewable, 36, 143, 169, 192, 194, 195, 201
Energy supply, 4, 20, 34, 59, 139
Energy use, 7, 27, 61, 90, 136, 140, 146, 148, 190, 201, 209, 242, 244
Energy use pattern, 228
Environmental planning management (EPM), 18
Environmental sustainability, 35, 71, 74, 93, 132, 179, 198, 248
Europe, 1, 24, 55, 62, 215
European countries, 62
European Union (EU), 20, 243

F
Feng shui, 122

G
Garden City model, 172
Governance, 25–27, 38, 39, 55, 71, 73, 85, 121, 150, 151, 218, 230–231, 239, 248, 249, 258–261
Green building certification, 62, 70, 96, 98, 100, 169, 174, 188, 190–191, 200–205, 208, 236
Green building council, 43
Green building development, 44, 68, 91, 92, 94, 99, 100, 169, 171, 174, 188, 189, 192, 210, 236, 257
Green building practice, 187–189
Green building standards, 94, 147, 174
Green corridors, 123, 134, 141
Green development, 9, 39, 115, 117, 120, 126, 129, 150, 215, 217, 220, 222, 228, 233, 236, 255, 258
Greenhouse gas emissions, 7, 14, 22, 40, 226, 237

Green industrial transformation, 117
Green infrastructure, 238, 246
Green Mark, 95, 191
Green roof, 62, 174, 227, 238
Green Star, 31, 65, 67
Green strategies, 123, 189, 219
Groundwater, 197
Guangming eco-city, 105, 125–131
Gubeikou Township, 168–170
Guiyang, 106, 115–120, 151, 227

H
Hangzhou, 201–204, 239
Hangzhou low carbon Science and Technology Museum (HLCSTM), 201–204
Harmonious relationship, 13, 232
Harmonisation, 22, 71, 92
Harmony, 117, 132, 190, 217, 232, 255
Health, 26, 27, 58, 72, 119, 121, 165, 227, 238, 246, 247
 urban, 23, 257
Healthy communities, 52, 53, 93, 228, 235
Healthy future, 17
Healthy planet, 53
High density, 34, 93, 144, 237
High-rise, 70, 110, 112, 122, 176, 178, 200, 242, 247
Historic perspective, 1–5
Hongqiao Business Park, 173–175
Human activities, 14, 18, 35
Human agency, 16
Human habitat, 16
Human settlements, 6, 16
Human–environment, 25
 relationships, 25
Humanity, 13, 83
Huzhou, 195–197
HVAC systems, 179, 180, 210

I

Indoor comfort, 203, 210, 224
Indoor environmental quality (IEQ), 30, 66, 69, 261
Industrial growth, 82
Industrial pattern, 119
Industrial products, 244, 245
Industrial revolution, 15
Industrial standard systems, 46
Industrial structure, 24, 113, 222
Industrialisation, 18, 44, 101, 116, 127
Information modelling techniques, 241–243
Infrastructure network, 4, 20, 226
Integrated design, 46, 129, 199, 232, 248
Integrated design approach (IDA), 198, 247
International cooperation, 76, 106, 114, 131, 143, 149
International norm, 262
Internet of Things (IoT), 140, 239, 259
Interplay, 7, 36, 38, 150, 256–259, 261, 263, 264

K

Key performance indicators (KPI), 95, 129, 135–137, 150, 177, 262
  breakdown, 134, 137

L

Leadership in Energy and Environmental Design (LEED), 31, 37, 60, 65, 67, 68, 95, 99, 163, 172, 174, 191, 200, 210, 216
  Neighbourhood Development (LEED-ND), 42, 68–70, 158, 159, 181
Lifecycle, 43, 46, 190, 241, 244, 261
Life cycle assessment (LCA), 28, 43, 46, 244, 245
Local initiative, 99–101, 106, 115, 149, 151
Localised, 31, 55, 75, 85, 222, 224
Low-carbon eco-city, 56, 95, 144, 249
Low-carbon industrial structure, 113
Low-carbon lifestyle, 113, 116, 119–120
Low carbon/low-carbon cities, 9, 16, 17, 21, 22, 24, 56, 93, 95, 112, 113, 124–126, 128, 129, 143, 146, 150, 161, 168, 169, 174, 175, 202, 215, 221, 222
Low-carbon transitions, 18, 162, 222, 264
Low-carbon transportation network, 174, 175
Low emission, 113, 140

M

Masterplan, 7, 35, 41, 69, 88, 109, 116, 121–123, 128, 134–136, 141, 150, 181
Meixi Lake Eco-City, 95, 105, 120–125
Ministry of Environmental Protection (MEP), 20, 88, 158
Ministry of Finance (MoF), 94, 168, 175
Ministry of Housing and Urban-Rural Development (MOHURD), 88, 91, 94–96, 135, 158, 168, 188, 189, 204, 223–225, 243
Misperceptions, 236
Mixed use, 34, 39, 52, 56, 57, 60, 62, 68, 70, 93, 134, 146, 171, 178, 180, 200, 208, 233
Multidimensional, 25, 247, 256
Multidisciplinary perspectives, 261

Multi-functional planning, 125
Multilateral, 18, 76
Multi-scalar, 10, 125, 211, 216, 217, 226–236, 247, 256
Multi-spatial, 27, 151

N
NanHai@COOL, 191, 204–207
National congress, 88, 89, 116, 255
National Development and Reform Commission (NDRC), 88, 90, 92, 126, 168, 202
National flagship project, 106, 120, 125, 149, 151, 230
National policy, 10, 18, 89
National target, 89, 90, 97
Natural environment, 14, 15, 24, 52, 53, 111, 121, 172, 190, 218, 233, 238, 246
Natural features, 24
Natural landscapes, 92, 238
Natural preservation, 16, 54, 71
Natural resource, 7, 13, 18, 29, 72, 117, 118, 132
Natural source, 29, 30
Natural system, 25
Natural ventilation, 46, 70, 174, 203, 206, 243
Nature-based solutions, 218, 245–247
Nearly zero energy building (NZEB), 223, 243, 244
Neighbourhood pattern, 31–34, 39, 68, 91, 180, 262
Neighbourhood rating systems, 69, 158
Neighbourhood sustainability, 39, 68
labelling, 69–70
New eastern city, 258
Ningbo, 99, 164–167, 191–195, 258
North America, 1, 70, 215

O
Operational energy, 138, 190

P
Passive design, 29, 61, 189–192, 195–196, 198, 200, 243
solar, 63
Passive energy sources, 179, 194
Passive house, 62–65, 95, 163, 179, 195–198, 223, 244
Passive house standard, 29, 54, 62, 95, 210, 216, 223
Passive measures, 29, 30, 209
Passive technologies, 29
Passivhaus, 195
Photovoltaic (pv), 61, 63, 90, 146, 190, 195, 203, 206, 210, 243
Pollution, vii, 4, 13, 14, 18, 20, 22, 27, 30, 53, 58, 70, 84, 91, 92, 96, 97, 113, 114, 119, 129, 136, 165, 172, 178, 179, 262
Public health, 246, 247

R
Rainwater, 35, 40, 59, 62, 70, 127, 139, 166, 177, 201, 207, 209, 220, 235
Rainwater harvesting, 45, 61
Renewables, 24, 30, 35–37, 40, 41, 45, 55, 58, 59, 63, 90, 91, 113, 119, 133, 136–140, 143, 146, 166, 169, 176, 181, 190, 192, 194, 195, 201, 203, 206, 209, 210, 216, 220, 221, 226, 236, 238–240, 243, 244, 259, 262
Resilience, 51, 217, 226–227, 246, 247
Resilient cities/city, 21, 238, 239

S

Sewage water, 59
Shanghai, 19, 54, 95, 96, 100, 106, 120, 131, 165, 167, 173–175, 188, 200–201, 240, 241, 243, 244, 247, 259
Shanghai Tower, 191, 200–201, 210
Shared bikes, 259, 260
Sharing services, 259
Shenzhen, 98, 105, 125, 191, 204–207, 259
Singapore, 95, 133, 136, 140, 142, 159–161, 171, 172
Sino-German Qingdao Eco-Industrial Park (SGQEP), 162–164, 173
Sino-Italian Ecological and Energy Efficient Building (SIEEB), 190, 198–200
Sino-Singaporean Guangzhou Knowledge City (SSGKC), 171–173
Sino-Singaporean Suzhou Industrial Park (SIP), 159–162, 171
Sino-Singaporean Tianjin Eco-City (SSTEC), 23–24, 112, 131–142, 171, 222, 249, 257
Sino-Swedish Wuxi Eco-City (SSWEC), 143–149
Small residential district (SRD), 93, 119, 159, 221, 233
Smart cities/city, 21, 139, 140, 151, 175, 219, 220, 239, 259
Social sustainability, 84
Solar system, 195
Spatial layout, 145
Spatial planning, 260
Stakeholder constellations, 106, 217, 229–232, 249, 264
(key) Stakeholders, 43, 44, 162, 198, 200, 204
Sustainability assessment, 28, 71–74
Sustainability evaluation system, 70
Sustainability framework, 21, 22
Sustainability indicator system, 258
Sustainability labelling, 69
Sustainability rating system, 68
Sustainability toolkit, 70
Sustainable cities, 9, 16–21, 60, 70–71, 120, 257
Sustainable technologies, 60, 61

T

Taihu New town, 144, 145, 149
Technical guidance, 258
Technological innovation, 15, 240
Tengtou Village, 164–167
Thermal bridging, 29, 63
Thermal comfort, 199, 210, 223, 224
Thermal insulation, 69, 163
Thermal mass, 29, 194
Thermal systems, 30
3D printing, 239–241, 256
3D-printed building, 240, 241
Topography, 28, 131
Traditional Chinese architecture, 233–235
Traditional cities, 15
Traditional culture, 13, 160
Traditional farming practice, 2
Traditional house, 234
Traditional planning practice, 34
Traditional principles, 30, 52, 89, 135, 232
Traditional water system, 25, 62, 131
Traditions, 2, 9, 13, 16, 20, 32, 34, 42, 45, 63, 88, 105, 113, 125, 129, 160, 170, 180, 217, 232–236, 238, 240, 241, 262
Traffic congestion, 20, 33, 92, 159, 234
Traffic management, 56, 163
Traffic system, 32
Transition
  energy, 264
  planning, 141

U

Underground parking, 179, 242
Underground space, 91
Underground storage, 62
Ur, 2–5
Urban density, 53, 70
Urban design, 23, 27, 35, 39, 56, 131, 216, 238
Urban efficiency, 256, 259
Urban form, 7, 16, 27, 34, 221, 233, 237, 238
Urban governance, 258–261
Urban growth, 10, 18, 84–85, 87, 132
Urban heat island (UHI), 227, 261
Urban infrastructure, 16, 33, 70, 143, 157
Urban poor, 7
Urban society, 9, 24–27, 95–98, 127, 224, 246

V

Ventilation strategies, 46, 192, 194
Vernacular, 75, 190, 236

W

Waste and pollution production, 18
Waste management, 23, 36, 112, 113, 139, 143, 148, 160, 263
Wastewater, 21, 132, 138, 160, 169, 170, 177, 179, 201, 257
Water catchment, 35, 56
Water consumption, 58, 119
Water efficiency, 58, 177
Water management, 21, 36, 55, 113, 114, 124, 126, 148, 238, 246
Water recycling, 36, 54, 146, 174, 179, 201, 259
Water supply, 35, 37, 40, 148
Water systems, 25, 61, 62, 109, 129, 169, 207, 209
Water treatment, 36, 61, 122, 160, 169, 170, 179, 235
Wellbeing, 43
Wind flow, 235
Wind intake, 243
Wind tunnel, 192, 200, 201
Wind turbines, 30, 195, 201, 204, 210
Window, 25, 29, 57, 64, 65, 88, 169, 177, 179, 202, 208, 209, 234, 235
Window sizing/size, 29, 63, 202
Window-to-wall ratio, 225

Y

Yingdong Ecological Village, 108, 111

Z

Zero energy building (ZEB), 223, 243–245
Zhejiang, 164–167, 191–198, 201–204, 240, 258

CPSIA information can be obtained
at www.ICGtesting.com
Printed in the USA
LVOW13*1735110518
576868LV00015B/406/P